The Imperative of Responsibility

The Imperative of Responsibility

In Search of an Ethics for the Technological Age

Hans Jonas

Translated by Hans Jonas with the Collaboration of David Herr

The University of Chicago Press
Chicago & London

Originally published as *Das Prinzip Verantwortung: Versuch einer Ethik für die technologische Zivilisation,* © Insel Verlag Frankfurt am Main, 1979; and *Macht oder Ohnmacht der Subjektivität? Das Leib-Seele-Problem im Vorfeld des Prinzips Verantwortung,* © Insel Verlag Frankfurt am Main, 1981.

The University of Chicago Press, Chicago 60637
The University of Chicago Press, Ltd., London

© 1984 by The University of Chicago
All rights reserved. Published 1984
Paperback Edition 1985
Printed in the United States of America
17 16 15 14 13 12 11   10 11 12 13 14

ISBN-13: 978-0-226-40597-1
ISBN-10: 0-226-40597-4

Library of Congress Cataloging-in-Publication Data

Jonas, Hans, 1903–
    The imperative of responsibility.

    Translation of: Das Prinzip Verantwortung; and
Macht oder Ohnmacht der Subjektivität?
    Includes index.
    1. Responsibility.   2. Technology and ethics.
3. Ethics.   I. Title.
BJ1453.J6613   1984   170'.42   83-18249

To my children
Ayalah, Jonathan, Gabrielle

Contents

Preface
To the English Edition

Modern technology, informed by an ever-deeper penetration of nature and propelled by the forces of market and politics, has enhanced human power beyond anything known or even dreamed of before. It is a power over matter, over life on earth, and over man himself; and it keeps growing at an accelerating pace. Its unfettered exercise for about two centuries now has raised the material estate of its wielders and main beneficiaries, the industrial "West," to heights equally unknown in the history of mankind. Not even the ravages of two world wars—themselves children of that overbrimming power—could slow the upward surge for long: it even gained from the spin-off of the hectic technological war effort in its aftermath. (The decades after World War II may well denote the high-water mark of technologic-economic ebullience.) But lately, the other side of the triumphal advance has begun to show its face, disturbing the euphoria of success with threats that are as novel as its welcomed fruits. Not counting the insanity of a sudden, suicidal atomic holocaust, which sane fear can avoid with relative ease, it is the slow, long-term, cumulative— the peaceful and constructive use of worldwide technological power, a use in which all of us collaborate as captive beneficiaries through rising production, consumption, and sheer population growth—that poses threats much harder to counter. The net total of these threats is the overtaxing of nature, environmental and (perhaps) human as well. Thresholds may be reached in one direction or another, points of no return, where processes initiated by us will run away from us on their own momentum— and toward disaster.

This (no longer novel) diagnosis forms the premise of the present work. In response, it develops the following major theses.

1. The altered, always enlarged nature of human action, with the magnitude and novelty of its works and their impact on man's global future,

raises moral issues for which past ethics, geared to the direct dealings of man with his fellowmen within narrow horizons of space and time, has left us unprepared. A new reflection on ethical principles—including such that, for lack of application, could hitherto remain silent—is required for coping with those issues.

2. The lengthened reach of our deeds moves *responsibility,* with no less than man's fate for its object, into the center of the ethical stage. Accordingly, a theory of responsibility, lacking so far, is set forth for both the private and the public sphere. Its axiom is that responsibility is a correlate of power and must be commensurate with the latter's scope and that of its exercise. For its discharge today, therefore, we need lengthened foresight, that is, a scientific futurology.

3. Even at its best, however, such an extrapolation from presently available data will always, in certainty and completeness of prediction, fall short of the causal pregnancy of our technological deeds. Consequently, an imaginative "heuristics of fear," replacing the former projections of hope, must tell us what is *possibly* at stake and what we must beware of. The magnitude of those stakes, taken together with the insufficiency of our predictive knowledge, leads to the pragmatic rule to give the prophecy of doom priority over the prophecy of bliss.

4. What we must avoid at all cost is determined by what we must preserve at all cost, and this in turn is predicated on the "image of man" we entertain. Formerly, this image was enshrined in the teachings of revealed religions. With their eclipse today, secular reason must base the normative concept of man on a cogent, at the least persuasive, doctrine of general being: metaphysics must underpin ethics. Hence, a speculative attempt is made at such an underpinning of man's duties toward himself, his distant posterity, and the plenitude of terrestrial life under his dominion. That attempt must brave the veto of reigning analytical theory against all attempts of this kind and indeed cannot hope for more than a tentative result. But dare it we must. A philosophy of nature is to bridge the alleged chasm between scientifically ascertainable "is" and morally binding "ought."

5. The thus gained conception of objective imperatives for man in the scheme of things enables us to discriminate between legitimate and illegitimate goal-settings to our Promethean power (a distinction encompassing but surpassing that between "realistic" and "unrealistic" goals). This discrimination is elaborated in assessing the potentials of "progress" up to the most ambitious idea of it—contemporary utopianism as represented by the Marxian alliance with technology. Against the immodesty of its goals, which maximize the inherent technological dangers of overstraining nature, the more modest and fitting goal is set to save the survival and humanity of man from the excesses of his own power.

This overview will, I hope, be of help to the reader, for whom things are not always made easy in this book (as they were not for the author either). When it first appeared in German, in 1979, it had been in the making for many years—ever since, in a 1959 lecture, "The Practical Uses of Theory" (now the Eighth Essay in my *The Phenomenon of Life*), I offered my first account of what I perceived to be the new relationship of theory and practice in our age and the role it plays in collective human affairs. Henceforth, my preoccupation with the impact of technology on the human condition grew apace with the growing and evermore obtrusive visibility of the phenomenon itself. As my thoughts progressed, they found public expression in a lengthening series of articles, and since the early 1970s these were from the outset meant to become parts of the comprehensive statement I had in mind, which would crystallize the single explorations into a whole as systematic as I could make it. The present volume is that statement—still fragmentary, of course, when measured against the daunting demands of the case. This qualification holds in particular where the argument passes from the description of facts and threats—by now fairly well established in a broadening consensus—to the true, prescriptive quest of the book: that after an *ethical* response to the facts. Here we must tread untrodden ground. Novel conditions and perils demand novel answers. A veritable *terra incognita* waits to be charted, and "practical reason" in the Kantian sense, that is, moral reason, is bidden to seek its way by its own uncertain light.

The original essays finally incorporated in this work are—in the order of chapters of which they now form parts—the following: (of Chapter 1) "Technology and Responsibility: Reflections on the New Tasks of Ethics," *Social Research* 40, no. 1 (1973): 31–54; (of Chapters 2, 5, 6) "Responsibility Today: The Ethics of an Endangered Future," *Social Research* 43, no. 1 (1976): 77–97; (of Chapter 4) "The Concept of Responsibility: An Inquiry into the Foundations of an Ethics for our Age," in H. T. Engelhardt and D. Callahan, eds., *Knowledge, Value, and Belief* (The Hastings Center: Hastings-on-Hudson, N.Y., 1977): 169–98; (of Chapters 4, 5, 6) "Reflections on Technology, Progress, and Utopia," *Social Research* 48, no. 3 (1981): 411–55; (of the Appendix) "Parallelism and Complementarity: The Psycho-Physical Problem in Spinoza and in the Succession of Niels Bohr," in R. Kennington, ed., *The Philosophy of Baruch Spinoza*, Studies in Philosophy and the History of Philosophy, vol. 7 (Washington, D.C.: Catholic University of America Press, 1980): 126–28; (of the Appendix) "On the Power or Impotence of Subjectivity," in S. F. Spicker and H. T. Engelhardt, eds., *Philosophical Dimensions of the Neuro-Medical Sciences* (Dordrecht and Boston: D. Reidel Publishing Co., 1976): 143–60 (copyright © 1976 by D. Reidel Publishing Company, Dordrecht, Holland). This material (sometimes altered) is reprinted here with permission.

Grateful acknowledgment is made to the National Endowment for the Humanities and to the Rockefeller Foundation for their generous grants that allowed me a year-long leave from my academic duties in which to begin the writing of this book. My very special thanks go to my co-fellows and friends of the Institute for Society, Ethics and the Life Sciences (Hastings Center), in whose congenial and stimulating company through more than a decade my ideas were nourished, tested, and refined.

New Rochelle, New York
August 1983   H. J.

1
The Altered Nature of
Human Action

All previous ethics—whether in the form of issuing direct enjoinders to do and not to do certain things, or in the form of defining principles for such enjoinders, or in the form of establishing the ground of obligation for obeying such principles—had these interconnected tacit premises in common: that the human condition, determined by the nature of man and the nature of things, was given once for all; that the human good on that basis was readily determinable; and that the range of human action and therefore responsibility was narrowly circumscribed. It will be the burden of the present argument to show that these premises no longer hold, and to reflect on the meaning of this fact for our moral condition. More specifically, it will be my contention that with certain developments of our powers the *nature of human action* has changed, and, since ethics is concerned with action, it should follow that the changed nature of human action calls for a change in ethics as well: this not merely in the sense that new objects of action have added to the case material on which received rules of conduct are to be applied, but in the more radical sense that the qualitatively novel nature of certain of our actions has opened up a whole new dimension of ethical relevance for which there is no precedent in the standards and canons of traditional ethics.

The novel powers I have in mind are, of course, those of modern *technology.* My first point, accordingly, is to ask how this technology affects the nature of our acting, in what ways it makes acting under its dominion *different* from what it has been through the ages. Since throughout those ages man was never without technology, the question involves the human difference of *modern* from previous technology.

I. The Example of Antiquity

Let us start with an ancient voice on man's powers and deeds which in an archetypal sense itself strikes, as it were, a technological note—the famous Chorus from Sophocles' *Antigone*.

Many the wonders but nothing more wondrous than man.
This thing crosses the sea in the winter's storm,
making his path through the roaring waves.
And she, the greatest of gods, the Earth—
deathless she is, and unwearied—he wears her away
as the ploughs go up and down from year to year
and his mules turn up the soil.

The tribes of the lighthearted birds he ensnares, and the races
of all the wild beasts and the salty brood of the sea,
with the twisted mesh of his nets, he leads captive, this clever man.
He controls with craft the beasts of the open air,
who roam the hills. The horse with his shaggy mane
he holds and harnesses, yoked about the neck,
and the strong bull of the mountain.

Speech and thought like the wind
and the feelings that make the town,
he has taught himself, and shelter against the cold,
refuge from rain. Ever resourceful is he.
He faces no future helpless. Only against death
shall he call for aid in vain. But from baffling maladies
has he contrived escape.

Clever beyond all dreams
the inventive craft that he has
which may drive him one time or another to well or ill.
When he honors the laws of the land and the gods' sworn right
high indeed is his city; but stateless the man
who dares to do what is shameful.

[Lines 335–370]

1. *Man and Nature*

This awestruck homage to man's powers tells of his violent and violating irruption into the cosmic order, the self-assertive invasion of nature's various domains by his restless cleverness; but also of his building—through the self-taught powers of speech and thought and social sentiment—the home for his very humanity, the artifact of the city. The raping of nature and the civilizing of man go hand in hand. Both are in defiance of the elements, the one by venturing into them and overpowering their creatures, the other by securing an enclave against them in the shelter of the city and its laws. Man is the maker of his life *qua* human, bending circumstances to his will and needs, and except against death he is never helpless.

Yet there is a subdued and even anxious quality about this appraisal of the marvel that is man, and nobody can mistake it for immodest bragging. Unspoken, but self-evident for those times, is the pervading knowledge behind it all that, for all his boundless resourcefulness, man is still small by the measure of the elements: precisely this makes his sallies into them so daring and allows those elements to tolerate his forwardness. Making free with the denizens of land and sea and air, he yet leaves the encompassing nature of those elements unchanged, and their generative powers undiminished. He cannot harm them by carving out his little kingdom from theirs. They last, while his schemes have their short-lived way. Much as he harries Earth, the greatest of gods, year after year with his plough—she is ageless and unwearied; her enduring patience he must and can trust, and to her cycle he must conform. And just as ageless is the sea. With all his netting of the salty brood, the spawning ocean is inexhaustible. Nor is it hurt by the plying of ships, nor sullied by what is jettisoned into its deeps. And no matter how many illnesses he contrives to cure, mortality does not bow to his cunning.

All this holds because before our time man's inroads into nature, as seen by himself, were essentially superficial and powerless to upset its appointed balance. (Hindsight reveals that they were not always so harmless in reality.) Nor is there a hint, in the *Antigone* chorus or anywhere else, that this is only a beginning and that greater things of artifice and power are yet to come—that man is embarked on an endless course of conquest. He had gone thus far in reducing necessity, had learned by his wits to wrest that much from it for the humanity of his life, and reflecting upon this, he was overcome by awe at his own boldness.

2. *The Man-Made Island of the "City"*

The room he has thus made was filled by the city of men—meant to enclose, and not to expand—and thereby a new balance was struck within the larger balance of the whole. All the good or ill to which man's inventive craft may drive him one time or another is inside the human enclave and does not touch the nature of things.

The immunity of the whole, untroubled in its depth by the importunities of man, that is, the essential immutability of Nature as the cosmic order, was indeed the backdrop to all of mortal man's enterprises, including his intrusions into that order itself. Man's life was played out between the abiding and the changing: the abiding was Nature, the changing his own works. The greatest of these works was the city, and on it he could confer some measure of abiding by the laws he made for it and undertook to honor. But no long-range certainty pertained to this contrived continuity. As a vulnerable artifact, the cultural construct can grow slack or go astray.

Not even within its artificial space, with all the freedom it gives to man's determination of self, can the arbitrary ever supersede the basic terms of his being. The very inconstancy of human fortunes assures the constancy of the human condition. Chance and luck and folly, the great equalizers in human affairs, act like an entropy of sorts and make all definite designs in the long run revert to the perennial norm. Cities rise and fall, rules come and go, families prosper and decline; no change is there to stay, and in the end, with all the temporary deflections balancing each other out, the state of man is as it always was. So here, too, in his very own artifact, the social world, man's control is small and his abiding nature prevails.

Still, this citadel of his own making, clearly set off from the rest of things and entrusted to him, was the whole and sole domain of man's responsible action. Nature was not an object of human responsibility—she taking care of herself and, with some coaxing and worrying, also of man: not ethics, only cleverness applied to her. But in the city, the social work of art, where men deal with men, cleverness must be wedded to morality, for this is the soul of its being. It is in this intrahuman frame, then, that all traditional ethics dwells, and it matches the size of action delimited by this frame.

II. Characteristics of Previous Ethics

Let us extract from the above those characteristics of human action which are relevant for a comparison with the state of things today.

1. All dealing with the nonhuman world, that is, the whole realm of *techne* (with the exception of medicine), was ethically neutral—in respect both of the object and the subject of such action: in respect of the object, because it impinged but little on the self-sustaining nature of things and thus raised no question of permanent injury to the integrity of its object, the natural order as a whole; and in respect of the agent subject it was ethically neutral because *techne* as an activity conceived itself as a determinate tribute to necessity and not as an indefinite, self-validating advance to mankind's major goal, claiming in its pursuit man's ultimate effort and concern. The real vocation of man lay elsewhere. In brief, action on nonhuman things did not constitute a sphere of authentic ethical significance.

2. Ethical significance belonged to the direct dealing of man with man, including the dealing with himself: all traditional ethics is *anthropocentric*.

3. For action in this domain, the entity "man" and his basic condition was considered constant in essence and not itself an object of reshaping *techne*.

4. The good and evil about which action had to care lay close to the act, either in the praxis itself or in its immediate reach, and were not

matters for remote planning. This proximity of ends pertained to time as well as space. The effective range of action was small, the time span of foresight, goal-setting, and accountability was short, control of circumstances limited. Proper conduct had its immediate criteria and almost immediate consummation. The long run of consequences beyond was left to chance, fate, or providence. Ethics accordingly was of the here and now, of occasions as they arise between men, of the recurrent, typical situations of private and public life. The good man was the one who met these contingencies with virtue and wisdom, cultivating these powers in himself, and for the rest resigning himself to the unknown.

All enjoinders and maxims of traditional ethics, materially different as they may be, show this confinement to the immediate setting of the action. "Love thy neighbor as thyself"; "Do unto others as you would wish them to do unto you"; "Instruct your child in the way of truth"; "Strive for excellence by developing and actualizing the best potentialities of your being *qua* man"; "Subordinate your individual good to the common good"; "Never treat your fellow man as a means only but always also as an end in himself"—and so on. Note that in all these maxims the agent and the "other" of his action are sharers of a common present. It is those who are alive now and in some relationship with me who have a claim on my conduct as it affects them by deed or omission. The ethical universe is composed of contemporaries, and its horizon to the future is confined by the foreseeable span of their lives. Similarly confined is its horizon of place, within which the agent and the other meet as neighbor, friend, or foe, as superior and subordinate, weaker and stronger, and in all the other roles in which humans interact with one another. To this proximate range of action all morality was geared.

It follows that the *knowledge* that is required—besides the moral will—to assure the morality of action fitted these limited terms: it was not the knowledge of the scientist or the expert, but knowledge of a kind readily available to all men of good will. Kant went so far as to say that "human reason can, in matters of morality, be easily brought to a high degree of accuracy and completeness even in the most ordinary intelligence";[1] that "there is no need of science or philosophy for knowing what man has to do in order to be honest and good, and indeed to be wise and virtuous. . . . [Ordinary intelligence] can have as good hope of hitting the mark as any philosopher can promise himself";[2] and again: "I need no elaborate acuteness to find out what I have to do so that my willing be morally good. Inexperienced regarding the course of the world, unable to anticipate all the contingencies that happen in it," I can yet know how to act in accordance with the moral law.[3]

Not every thinker in ethics, it is true, went so far in discounting the cognitive side of moral action. But even when it received much greater emphasis, as in Aristotle, where the discernment of the situation and what is fitting for it makes considerable demands on experience and judgment, such knowledge has nothing to do with the science of things. It implies, of course, a general conception of the human good as such, a conception predicated on the presumed invariables of man's nature and condition, which may or may not find expression in a theory of its own. But its translation into practice requires a knowledge of the here and now, and this is entirely nontheoretical. This "knowledge" proper to virtue (of the "where, when, to whom, and how") stays with the immediate issue, in whose defined context the action as the agent's own takes its course and within which it terminates. The good or bad of the action is wholly decided within that short-term context. Its authorship is unquestioned, and its moral quality shines forth from it, visible to its witnesses. No one was held responsible for the unintended later effects of his well-intentioned, well-considered, and well-performed act. The short arm of human power did not call for a long arm of predictive knowledge; the shortness of the one is as little culpable as that of the other. Precisely because the human good, known in its generality, is the same for all time, its realization or violation takes place at each time, and its complete locus is always the present.

III. New Dimensions of Responsibility

All this has decisively changed. Modern technology has introduced actions of such novel scale, objects, and consequences that the framework of former ethics can no longer contain them. The *Antigone* chorus on the *deinotes*, the wondrous power, of man would have to read differently now; and its admonition to the individual to honor the laws of the land would no longer be enough. The gods, too, whose venerable right could check the headlong rush of human action, are long gone. To be sure, the old prescriptions of the "neighbor" ethics—of justice, charity, honesty, and so on—still hold in their intimate immediacy for the nearest, day-by-day sphere of human interaction. But this sphere is overshadowed by a growing realm of collective action where doer, deed, and effect are no longer the same as they were in the proximate sphere, and which by the enormity of its powers forces upon ethics a new dimension of responsibility never dreamed of before.

1. *The Vulnerability of Nature*

Take, for instance, as the first major change in the inherited picture, the critical vulnerability of nature to man's technological intervention—unsuspected before it began to show itself in damage already done. This discovery, whose shock led to the concept and nascent science of ecology, alters the very concept of ourselves as a causal agency in the larger scheme of things. It brings to light, through the effects, that the nature of human action has *de facto* changed, and that an object of an entirely new order—no less than the whole biosphere of the planet—has been added to what we must be responsible for because of our power over it. And of what surpassing importance an object, dwarfing all previous objects of active man! Nature as a human responsibility is surely a novum to be pondered in ethical theory. What kind of obligation is operative in it? Is it more than a utilitarian concern? Is it just prudence that bids us not to kill the goose that lays the golden eggs, or saw off the branch on which we sit? But the "we" who sit here and who may fall into the abyss—who is it? And what is *my* interest in its sitting or falling?

Insofar as it is the fate of *man,* as affected by the condition of nature, which makes our concern about the preservation of nature a *moral* concern, such concern admittedly still retains the anthropocentric focus of all classical ethics. Even so, the difference is great. The containment of nearness and contemporaneity is gone, swept away by the spatial spread and time span of the cause-effect trains which technological practice sets afoot, even when undertaken for proximate ends. Their irreversibility conjoined to their aggregate magnitude injects another novel factor into the moral equation. Add to this their cumulative character: their effects keep adding themselves to one another, with the result that the situation for later subjects and their choices of action will be progressively different from that of the initial agent and ever more the fated product of what was done before. All traditional ethics reckoned only with noncumulative behavior.[4] The basic situation between persons, where virtue must prove and vice expose itself, remains always the same, and every deed begins afresh from this basis. The recurring occasions which pose their appropriate alternatives for human conduct—courage or cowardice, moderation or excess, truth or mendacity, and so on—each time reinstate the primordial conditions from which action takes off. These were never superseded, and thus moral actions were largely "typical," that is, conforming to precedent. In contrast with this, the cumulative self-propagation of the technological change of the world constantly overtakes the conditions of its contributing acts and moves through none but unprecedented situations, for which the lessons of experience are powerless. And not even content with changing its beginning to the point of unrecognizability, the cumulation as such may consume the basis of the whole series, the very condition of itself. All this would have to be cointended in the will of the single action if this is to be a morally responsible one.

2. *The New Role of Knowledge in Morality*

Knowledge, under these circumstances, becomes a prime duty beyond anything claimed for it heretofore, and the knowledge must be commensurate with the causal scale of our action. The fact that it cannot really be thus commensurate, that is, that the predictive knowledge falls behind the technical knowledge that nourishes our power to act, itself assumes ethical importance. The gap between the ability to foretell and the power to act creates a novel moral problem. With the latter so superior to the former, recognition of ignorance becomes the obverse of the duty to know and thus part of the ethics that must govern the evermore necessary self-policing of our outsized might. No previous ethics had to consider the global condition of human life and the far-off future, even existence, of the race. These now being an issue demands, in brief, a new conception of duties and rights, for which previous ethics and metaphysics provide not even the principles, let alone a ready doctrine.

3. Has Nature "Rights" Also?

And what if the new kind of human action would mean that more than the interest of man alone is to be considered—that our duty extends farther, and the anthropocentric confinement of former ethics no longer holds? It is at least not senseless anymore to ask whether the condition of extrahuman nature, the biosphere as a whole and in its parts, now subject to our power, has become a human trust and has something of a moral claim on us not only for our ulterior sake but for its own and in its own right. If this were the case it would require quite some rethinking in basic principles of ethics. It would mean to seek not only the human good but also the good of things extrahuman, that is, to extend the recognition of "ends in themselves" beyond the sphere of man and make the human good include the care for them. No previous ethics (outside of religion) has prepared us for such a role of stewardship—and the dominant, scientific view of *Nature* has prepared us even less. Indeed, that view emphatically denies us all conceptual means to think of Nature as something to be honored, having reduced it to the indifference of necessity and accident, and divested it of any dignity of ends. But still, a silent plea for sparing its integrity seems to issue from the threatened plenitude of the living world. Should we heed this plea, should we recognize its claim as morally binding because sanctioned by the nature of things, or dismiss it as a mere sentiment on our part, which we may indulge as far as we wish and can afford to do? If the former, it would (if taken seriously in its theoretical implications) push the necessary rethinking beyond the doctrine of action, that is, ethics, into the doctrine of being, that is, metaphysics, in which all ethics must ultimately be grounded. On this speculative subject I will say no more here than that we should keep ourselves open to the thought that natural science may not tell the whole story about Nature.

IV. Technology as the "Calling" of Mankind

1. *Homo Faber* over *Homo Sapiens*

Returning to strictly intrahuman considerations, there is another ethical aspect to the growth of *techne* as a pursuit beyond the pragmatically limited terms of former times. Then, so we found, *techne* was a measured tribute to necessity, not the road to mankind's chosen goal—a means with a finite measure of adequacy to well-defined proximate ends. Now, *techne* in the form of modern technology has turned into an infinite forward-thrust of the race, its most significant enterprise, in whose permanent, self-transcending advance to ever greater things the vocation of man tends to be seen, and whose success of maximal control over things and himself appears as the consummation of his destiny. Thus the triumph of *homo faber* over his external object means also his triumph in the internal constitution of *homo sapiens,* of whom he used to be a subsidiary part. In other words, technology, apart from its objective works, assumes ethical significance by the central place it now occupies in human purpose. Its cumulative creation, the expanding artificial environment, continuously reinforces the particular powers in man that created it, by compelling their unceasing inventive employment in its management and further advance, and by rewarding them with additional success—which only adds to the relentless claim. This positive feedback of functional necessity and reward—in whose dynamics pride of achievement must not be forgotten—assures the growing ascendancy of one side of man's nature over all the others, and inevitably at their expense. If nothing succeeds like success, nothing also entraps like success. Outshining in prestige and starving in resources whatever else belongs to the fullness of man, the expansion of his power is accompanied by a contraction of his self-conception and being. In the image he entertains of himself—the programmatic idea which determines his actual being as much as it reflects it—man now is evermore the maker of what he has made and the doer of what he can do, and most of all the preparer of what he will be able to do next. But who is "he"? Not you or I: it is the aggregate, not the individual doer or deed that matters here; and the indefinite future, rather than the contemporary context of the action, constitutes the relevant horizon of responsibility. This requires imperatives of a new sort. If the realm of making has invaded the space of essential action, then morality must invade the realm of making, from which it has formerly stayed aloof, and must do so in the form of public policy. Public policy has never had to deal before with issues of such inclusiveness and such lengths of anticipation. In fact, the changed nature of human action changes the very nature of politics.

2. *The Universal City as a Second Nature*

For the boundary between "city" and "nature" has been obliterated: the city of men, once an enclave in the nonhuman world, spreads over the whole of terrestrial nature and usurps its place. The difference between the artificial and the natural has vanished, the natural is swallowed up in the sphere of the artificial, and at the same time the total artifact (the works of man that have become "the world" and as such envelop their makers) generates a "nature" of its own, that is, a necessity with which human freedom has to cope in an entirely new sense.

Once it could be said *Fiat justitia, pereat mundus,* "Let justice be done, and may the world perish"—where "world," of course, meant the renewable enclave in the imperishable whole. Not even rhetorically can the like be said anymore when the perishing of the whole through the doings of man—be they just or unjust—has become a real possibility. Issues never legislated come into the purview of the laws which the total city must give itself so that there will be a world for the generations of man to come.

3. *Man's Presence in the World as an Imperative*

That there *ought* to be through all future time such a world fit for human habitation, and that it ought in all future time to be inhabited by a mankind worthy of the human name, will be readily affirmed as a general axiom or a persuasive desirability of speculative imagination (as persuasive and as undemonstrable as the proposition that there being a world at all is "better" than there being none): but as a *moral* proposition, namely, a practical *obligation* toward the posterity of a distant future, and a principle of decision in present action, it is quite different from the imperatives of the previous ethics of contemporaneity; and it has entered the moral scene only with our novel powers and range of prescience.

The *presence of man in the world* had been a first and unquestionable given, from which all idea of obligation in human conduct started out. Now it has itself become an *object* of obligation: the obligation namely to ensure the very premise of all obligation, that is, the *foothold* for a moral universe in the physical world—the existence of mere *candidates* for a moral order. This entails, among other things, the duty to preserve this physical world in such a state that the conditions for that presence remain intact; which in turn means protecting the world's vulnerability from what could imperil those very conditions. The difference this makes for ethics may be illustrated in one example.

V. Old and New Imperatives

1. Kant's categorical imperative said: "Act so that you *can* will that the maxim of your action be made the principle of a universal law." The "can"

here invoked is that of reason and its consistency with itself: *Given* the existence of a community of human agents (acting rational beings), the action must be such that it can without self-contradiction be imagined as a general practice of that community. Mark that the basic reflection of morals here is not itself a moral but a logical one: The "I *can* will" or "I *cannot* will" expresses logical compatibility or incompatibility, not moral approbation or revulsion. But there is no self-contradiction in the thought that humanity would once come to an end, therefore also none in the thought that the happiness of present and proximate generations would be bought with the unhappiness or even nonexistence of later ones—as little as, after all, in the inverse thought that the existence or happiness of later generations would be bought with the unhappiness or even partial extinction of present ones. The sacrifice of the future for the present is *logically* no more open to attack than the sacrifice of the present for the future. The difference is only that in the one case the series goes on, and in the other it does not (or: its future ending is contemplated). But that it *ought to go on,* regardless of the distribution of happiness or unhappiness, even with a persistent preponderance of unhappiness over happiness, nay, of immorality over morality[5]—this cannot be derived from the rule of self-consistency *within* the series, long or short as it happens to be: it is a commandment of a very different kind, lying outside and "prior" to the series as a whole, and its ultimate grounding can only be metaphysical.

2. An imperative responding to the new type of human action and addressed to the new type of agency that operates it might run thus: "Act so that the effects of your action are compatible with the permanence of genuine human life"; or expressed negatively: "Act so that the effects of your action are not destructive of the future possibility of such life"; or simply: "Do not compromise the conditions for an indefinite continuation of humanity on earth"; or, again turned positive: "In your present choices, include the future wholeness of Man among the objects of your will."

3. It is immediately obvious that no rational contradiction is involved in the violation of this kind of imperative. I can will the present good with sacrifice of the future good. Just as I can will my own end, I can will that of humanity. Without falling into contradiction with myself, I can prefer a short fireworks display of the most extreme "self-fulfillment," for myself or for the world, to the boredom of an endless continuation in mediocrity.

However, the new imperative says precisely that we may risk our own life—but not that of humanity; and that Achilles indeed had the right to choose for himself a short life of glorious deeds over a long life of inglorious security (with the tacit premise that a posterity would be there to know and tell of his deeds), but that we do not have the right to choose, or even risk, nonexistence for future generations on account of a better life for the present one. Why we do not have this right, why on the contrary we have an obligation toward that which does not yet exist and never need exist at all—an obligation not only toward its fortunes in case it happens to exist, but toward its coming to exist in the first place, to which as nonexistent "it" surely has no claim: to underpin this proposition theoretically is by no means easy and without religion perhaps impossible. At present, our imperative simply posits it without proof, as an axiom.

4. It is also evident that the new imperative addresses itself to public policy rather than private conduct, which is not in the causal dimension to which that imperative applies. Kant's categorical imperative was addressed to the individual, and its criterion was instantaneous. It enjoined each of us to consider what would happen *if* the *maxim* of my present action were made, or at this moment already were, the principle of a universal legislation; the self-consistency or inconsistency of such a *hypothetical* universalization is made the test for my *private* choice. But it was no part of the reasoning that there is any probability of my private choice in fact becoming universal law, or that it might contribute to its becoming that. Indeed, *real* consequences are not considered at all, and the principle is one not of objective responsibility but of the subjective quality of my self-determination. The new imperative invokes a different consistency: not that of the act with itself, but that of its eventual *effects* with the continuance of human agency in times to come. And the "universalization" it contemplates is by no means hypothetical—that is, a purely logical transference from the individual "me" to an imaginary, causally unrelated "all" ("*if* everybody acted like that"); on the contrary, the actions subject to the new imperative—actions of the collective whole—have their universal reference in their actual scope of efficacy: they "totalize" themselves in the progress of their momentum and thus are bound to terminate in shaping the universal dispensation of things. This adds a *time* horizon to the moral calculus which is entirely absent from the instantaneous logical operation of the Kantian imperative: whereas the latter extrapolates into an ever-present order of abstract compatibility, our imperative extrapolates into a predictable real *future* as the open-ended dimension of our responsibility.

VI. Earlier Forms of "Future-oriented Ethics"

Now it may be objected that with Kant we have chosen an extreme example of the ethics of subjective intention (*Gesinnungsethik*), and that our assertion of the present-oriented character of all former ethics, as holding among contemporaries, is contradicted by several ethical forms of the past. The following three examples come to mind: the conduct of earthly life (to the point of sacrificing its entire happiness) with a view to the eternal salvation of the soul; the long-range concern of the legislator and statesman for the future common weal; and the politics of utopia, with its readiness to use those living now as a mere means to a goal that lies in a future after their time, or to exterminate them as obstacles in its way—of which revolutionary Marxism is the prime example.

1. *The Ethics of Fulfillment in the Life Hereafter*

Of these three cases the first and third share the trait of placing the future above the present as the possible locus of absolute value, thus demoting the present to a mere preparation for the future. An important difference is that in the religious case the acting down here is not credited with bringing on the future bliss by its own causality (as revolutionary action is supposed to do), but is merely supposed to *qualify* the agent for it, namely, in the eyes of God, to whom faith must entrust its realization. That qualification, however, consists in a life pleasing to God, of which in general it may be assumed that it is the best, most worthwhile life in itself anyway, thus worthy to be chosen for its own sake and not merely for that of eventual future bliss. Indeed, when chosen mainly from that reward motive, the life in question would lose in worth and therewith even in its qualifying strength. That is to say, the latter is the greater, the less intended it is. When we then ask what human qualities are held to procure the qualification, that is, to constitute a life pleasing to God, we must look at the life prescriptions of the particular creeds—and these we may often find to be just those prescriptions of justice, charity, purity of heart, etc., which would, or could, be prescribed by an innerworldly ethic of the classical sort as well. Thus in the "moderate" version of the belief in the soul's salvation (of which, if I am not mistaken, Judaism is an example) we still deal, after all, with an ethics of contemporaneity and immediacy, notwithstanding the transcendent goal; and what ethics it might concretely be in this or that historical case—that is not deducible from the transcendent goal as such (of whose content no idea can be formed anyway), but is told by the way in which the "life pleasing to God," said to be the precondition for it, was in each instance given material content.

It may happen, however, that the content is such—and this is the case in the "extreme" forms of the soul salvation doctrine—that its practice, that is, the fulfillment of the "precondition," can in no way be regarded as of value in itself but is merely the stake in a wager, with whose loss, that is, the failure to attain the eternal reward, all would be lost. For in this case of the dreadful metaphysical bet as elaborated by Pascal, the stake is one's entire earthly existence with all its possibilities of enjoyment and fulfillment, whose very renunciation is made the price of eternal salvation. In this category belong all those forms of radical mortification of the flesh, of life-denying asceticism, whose practitioners would have cheated themselves out of everything if their expectations were disappointed. This otherworldly wager differs from the calculus of ordinary, this-worldly hedonism, with its considered risks of sometime-renunciations and de-

ferments, merely by the totality of its *quid pro quo* and the surpassing
nature of the chance for which the stakes are risked. But just this surpassing expectation moves the whole undertaking out of the realm of
ethics. Between the finite and the infinite, the temporal and the eternal,
there is no commensurability and thus no meaningful comparison; that
is, there is neither a qualitative nor a quantitative sense in which one is
preferable to the other. Concerning the *value* of the goal, whose informed
appraisal ought to form an essential element of *ethical* decision, there is
nothing but the empty assertion that it is the ultimate value. Also lacking
is the *causal* relation—which at least *ethical* thinking requires—between
the action and its (hoped-for) result; that "result," so we saw, is conceived
not as being effected by present renunciation but merely as promised from
elsewhere in compensation for it.

If one inquires *why* the this-worldly renunciation is considered so mer-
itorious that it may dare to expect this kind of indemnification or reward,
one answer might be that the flesh is sinful, desire is evil, and the world
is impure. In this case (as in the somewhat different case where individuation as such is regarded as bad) asceticism does represent, after all, a
genuine instrumentality of action and a path to internal goal-achievement
through one's own performance: the path, namely, from impurity to purity,
from sinfulness to sanctity, from bondage to freedom, from selfhood to
self-transcendence. Insofar as it is such a "path," asceticism is already
in itself the *best* sort of life by the metaphysical criteria assumed. But in
this case we are dealing again with an ethic of the here and now: a form—
albeit a supremely egotistic and individualistic form—of the ethic of self-
perfection, whose inward exertions may indeed attain to those peak moments of spiritual illumination, which are a present foretaste of the future
reward: a mystical experience of the Absolute.

In sum, we can say that, insofar as this whole complex of otherworldly
striving falls within ethics at all (as do, for instance, the aforementioned
"moderate" forms in which a life good in itself forms the condition for
eternal reward), it too fits our thesis concerning the orientation of all
previous ethics to the present.

2. *The Statesman's Responsibility for the Future*

What about the examples of *innerworldly* future-oriented ethics, which
alone do really belong to rational ethics in that they reckon with a known
cause-effect pattern? We mentioned in the second place the long-range
care of the legislator and statesman for the future good of the commonwealth. Greek political theory is on the whole silent about the *time* aspect
which interests us here; but this silence itself is revealing. Something can
be gathered from the praise of great lawgivers like Solon and Lycurgus
or from the censure of a statesman like Pericles. The praise of the lawgiver
includes, it is true, the durability of his creation, but not his planning

ahead of something that is to come about only in aftertimes and not attainable already to his contemporaries. His endeavor is to create a viable political structure, and the test of viability is in the enduring of his creation—a changeless enduring if possible. The best state, so it was thought, is also the best for the future, precisely because the stable equilibrium of its present ensures its future as such; and it will then, of course, be the best state *in* that future as well, since the criteria of a good order (of which durability is one) do not change. They do not change because human nature does not change, which with its imperfections is included in the conception which the wise lawgiver must have of a viable political order. This conception thus aims not at the ideally perfect state but rather at the realistically best, that is, the best possible state—and this is now just as possible, and just as imperiled, as it will always be. But this very peril, which threatens all order with the disorder of the human passions, makes necessary, in addition to the singular, founding wisdom of the lawgiver, the continuous, governing wisdom of the statesman. The reproach of Socrates against the politics of Pericles, be it noted, is not that, in the end after his death, his grandiose schemes came to nought, but rather that with such grandiose schemes (including their initial successes) he had already in his own time turned the Athenians' heads and corrupted their civic virtues. Athens' current misfortune thus was blamed not on the eventual failure of those policies but on the blemish at their roots, which even "success" in their own terms would not have made better in retrospect. What would have been good at that time would be that still today and would most probably have survived into the present.

The foresight of the statesman thus consists in the wisdom and moderation he devotes to the *present*. This present is not here for the sake of a future different from (and superior to) it in type, but rather proves itself—luck permitting—in a future still like itself, and so must be as justified already in itself as its succession is hoped to be. Duration, in short, results as a concomitant of what is good now and at all times. Certainly, political action has a wider time span of effect and responsibility than private action, but its ethics, according to the premodern view, is still none other than the present-oriented one, applied to a life form of longer duration.

3. *The Modern Utopia*

a) This changes only with what, in my third example, I called the politics of utopia, which is a thoroughly modern phenomenon and presupposes a previously unknown, dynamic eschatology of history. The religious eschatologies of earlier times do not yet represent this case, although they prepare for it. Messianism, for example, does not ordain a messianic politics, but leaves the coming of the Messiah to divine dispensation. Human behavior is implicated in it only in the sense that it can make itself worthy of the event through fulfilling those very norms to which it is subject even without such a prospect. Here we find to hold on the collective scale what we previously found to hold on the personal scale with regard to otherworldly hopes: the here and now is certainly overarched by them, but is not entrusted with their active realization. It serves them the better, the more faithful it remains to its own God-given law, whose fulfillment lies entirely within itself.

b) Here, too, there did occur the extreme form, where the "urgers of the end" took matters into their own hands and with one last thrust of earthly action tried to bring about the messianic kingdom or millennium, for which they considered the time ripe. In fact, some of the chiliastic movements, especially at the beginning of the modern era, lead into the neighborhood of utopian politics, particularly when they are not content with merely having made a start and clearing the path, but when they make a positive beginning with the Kingdom of God, of whose *contents* they have a definite conception. Insofar as ideas of social equality and justice play a role in this conception, the characteristic motivation of modern utopian ethics is already there: but not yet the yawning gulf, stretching across generations, between now and later, means and end, action and goal, which marks the modern, secularized eschatology, that is, modern political utopianism. It is still an ethic of the self-vindicating present, not of the retroactively vindicating future: the true man is already there, and even, in the "community of the saints," the kingdom of God from the moment they realize it in their own midst, as ordained and held to be possible in the dawning fulness of time. The assault, however, against the establishments of the world that still oppose its spreading, is made in the expectation of a Jericho-like miracle, not as a mediated process of historical causation. The last step to the innerworldly utopian ethic of history is yet to be taken.

c) Only with the advent of modern *progress,* both as a fact and as an idea, did the possibility emerge of conceiving everything past as a stepping-stone to the present and of everything present as a stepping-stone to the future. When this notion (which in itself, as unlimited, distinguishes no stage as final and leaves to each the immediacy of its own present) is wed with a secularized eschatology which assigns to the absolute, defined in terms of this world, a finite place in time, and when to this is added a conception of a teleological dynamism which leads to the final state of affairs—then we have the conceptual prerequisites for a utopian politics. "To found the kingdom of heaven already upon earth" (Heinrich Heine) presupposes some idea of what such an earthly kingdom of heaven would look like (or so one would think—but on this point the theory displays a remarkable blank). In any case, even lacking such an idea, the resolute secular eschatology entails a conception of human events that radically demotes to provisional status all that goes before, stripping it of its independent validity and at best making it the vehicle for reaching the promised state of things that is yet to come—a means to the future end which alone is worthy in itself.

Here in fact is a break with the past, and what we have said concerning the present-oriented character of all previous ethics and their common premise of the persistence of human nature is no longer true of the teaching which represents this break most clearly, the Marxist philosophy of history and its corresponding ethic of action. Action takes place for the sake of a future which neither the agent nor the victim nor their contemporaries will live to enjoy. The obligations upon the now issue from that goal, not from the good and ill of the contemporary world; and the norms of action are just as provisional, indeed just as "inauthentic," as the conditions which it will transmute into the higher state. The ethic of revolutionary eschatology considers itself an ethic of transition, while the consummate, true ethic (essentially still unknown) will only come into its own after the harsh interim morality (which can last a long time) has created the conditions for it and thereby abrogated itself.

Thus there already exists, in Marxism, a future-oriented ethic, with a distance of vision, a time span of affirmed responsibility, a scope of object (= all of future humanity), and a depth of concern (the whole future nature of man)—and, as we might already add, with a sense for the powers of technology—which in all these respects stands comparison with the ethic for which we want to plead here. All the more important it is to determine the relation between these two ethical positions which, as answers to the unprecedented modern situation and especially to its technology, have so much in common over against premodern ethics and yet are so different from one another. This must wait until we have heard more about the problems and tasks which the ethic here envisaged has to deal with, and which are posed by the colossal progress of technology. For technology's power over human destiny has overtaken even that of communism, which no less than capitalism thought merely to make use of it. We say this much in advance: while both positions concern themselves with the utopian possibilities of this technology, the ethic we are looking for is *not* eschatological and, in a sense yet to be specified, is anti-utopian.

VII. Man as an Object of Technology

Our comparison dealt with the historical forms of the ethics of contemporaneity and immediacy, for which the Kantian case served only as an example. What stands in question is not their validity within their own frame of reference but their sufficiency for those new dimensions of human action which transcend that frame. Our thesis is that the new kinds and dimensions of action require a commensurate ethic of foresight and responsibility which is as novel as the eventualities which it must meet. We have seen that these are the eventualities that arise out of the works of *homo faber* in the era of technology. But among those novel works we have not mentioned yet the potentially most ominous class. We have considered *techne* only as applied to the nonhuman realm. But man himself has been added to the objects of technology. *Homo faber* is turning upon himself and gets ready to make over the maker of all the rest. This consummation of his power, which may well portend the overpowering of man, this final imposition of art on nature, calls upon the utter resources of ethical thought, which never before has been faced with elective alternatives to what were considered the definite terms of the human condition.

1. *Extension of Life Span*

Take, for instance, the most basic of these "givens," man's *mortality*. Who ever before had to make up his mind on its desirable and *eligible* measure? There was nothing to choose about the upper limit, the "three-score years and ten, or by reason of strength fourscore." Its inexorable rule was the subject of lament, submission, or vain (not to say foolish) wish-dreams about possible exceptions—strangely enough, almost never of affirmation. The intellectual imagination of a George Bernard Shaw and a Jonathan Swift speculated on the privilege of not having to die, or the curse of not being able to die. (Swift with the latter was the more perspicacious of the two.) Myth and legend toyed with such themes against the acknowledged background of the unalterable, which made the earnest man rather pray "teach us to number our days that we may get a heart of wisdom" (Psalm 90). Nothing of this was in the realm of doing and effective decision. The question was only how to relate to the stubborn fact.

But lately the dark cloud of inevitability seems to lift. A practical hope is held out by certain advances in cell biology to prolong, perhaps indefinitely extend, the span of life by counteracting biochemical processes of aging. Death no longer appears as a necessity belonging to the nature of life, but as an avoidable, at least in principle tractable and long-delayable, organic malfunction. A perennial yearning of mortal man seems to come nearer fulfillment. And for the first time we have in earnest to ask the questions "How desirable is this? How desirable for the individual, and how for the species?" These questions involve the very meaning of our finitude, the attitude toward death, and the general biological significance of the balance of death and procreation. Even prior to such ultimate questions are the more pragmatic ones of who should be eligible for the boon: Persons of particular quality and merit? Of social eminence? Those who can pay for it? Everybody? The last would seem the only just course. But it would have to be paid for at the opposite end, at the source. For clearly, on a population-wide scale, the price of extended age must be a proportional slowing of replacement, that is, a diminished access of new life. The result would be a decreasing proportion of youth in an increasingly aged population. How good or bad would that be for the general condition of man? Would the species gain or lose? And how *right* would it be to preempt the place of youth? Having to die is bound up with having been born: mortality is but the other side of the perennial spring of "natality" (to use Hannah Arendt's term). This had always been ordained; now its meaning has to be pondered in the sphere of decision.

To take the extreme (not that it will ever be obtained): if we abolish death, we must abolish procreation as well, for the latter is life's answer to the former, and so we would have a world of old age with no youth, and of known individuals with no surprises of such that had never been before. But this perhaps is precisely the wisdom in the harsh dispensation of our mortality: that it grants us the eternally renewed promise of the freshness, immediacy, and eagerness of youth, together with the supply of otherness as such. There is no substitute for this in the greater accumulation of prolonged experience: it can never recapture the unique privilege of seeing the world for the first time and with new eyes; never relive the wonder which, according to Plato, is the beginning of philosophy; never the curiosity of the child, which rarely enough lives on as thirst for knowledge in the adult, until it wanes there too. This ever renewed beginning, which is only to be had at the price of ever repeated ending, may well be mankind's hope, its safeguard against lapsing into boredom and routine, its chance of retaining the spontaneity of life. Also, the role of the *memento mori* in the individual's life must be considered, and what its attenuation to indefiniteness may do to it. Perhaps a nonnegotiable limit to our expected time is necessary for each of us as the incentive to number our days and make them count.

So it could be that what by intent is a philanthropic gift of science to man, the partial granting of his oldest wish—to escape the curse of mortality—turns out to be to the detriment of man. I am not indulging in prediction and, in spite of my noticeable bias, not even in valuation. My point is that already the promised gift raises questions that had never to be asked before in terms of practical choice, and that no principle of former ethics, which took the human constants for granted, is competent to deal with them. And yet they must be dealt with ethically and by principle and not merely by the pressure of interests.

2. *Behavior Control*

It is similar with all the other, quasi-utopian possibilities which progress in the biomedical sciences has partly already placed at our disposal and partly holds in prospect for eventual translation into technological know-how. Of these, *behavior control* is much nearer to practical readiness than the still hypothetical prospect I have just been discussing, and the ethical questions it raises are less profound but have a more direct bearing on the moral conception of man. Here again, the new kind of intervention exceeds the old ethical categories. They have not equipped us to rule, for example, on mental control by chemical means or by direct electrical action on the brain via implanted electrodes—undertaken, let us assume, for defensible and even laudable ends. The mixture of beneficial and dangerous potentials is obvious, but the lines are not easy to draw. Relief of mental patients from distressing and disabling symptoms seems un-equivocally beneficial. But from the relief of the *patient,* a goal entirely in the tradition of the medical art, there is an easy passage to the relief of *society* from the inconvenience of difficult individual behavior among its members: that is, the passage from medical to social application; and this opens up an indefinite field with grave potentials. The troublesome problems of rule and unruliness in modern mass society make the exten-sion of such control methods to nonmedical categories extremely tempting for social management. Numerous questions of human rights and dignity arise. The difficult question of preempting versus enabling care insists on concrete answers. Shall we induce learning attitudes in schoolchildren by the mass administration of drugs, circumventing the appeal to autonomous motivation? Shall we overcome aggression by electronic pacification of brain areas? Shall we generate sensations of happiness or pleasure or at least contentment through independent stimulation (or tranquilizing) of the appropriate centers—independent, that is, of the objects of happiness, pleasure, or content and their attainment in personal living and achieving? Candidacies could be multiplied. Business firms might become interested in some of these techniques for performance increase among their employees.

Regardless of the question of compulsion or consent, and regardless also of the question of undesirable side-effects, each time we thus bypass the human way of dealing with human problems, short-circuiting it by an impersonal mechanism, we have taken away something from the dignity of personal selfhood and advanced a further step on the road from re-sponsible subjects to programmed behavior systems. Social functionalism, important as it is, is only one side of the question. Decisive is the question of what kind of individuals the society is composed of—to make its ex-istence valuable as a whole. Somewhere along the line of increasing social manageability at the price of individual autonomy, the question of the worthwhileness of the whole human enterprise must pose itself. An-swering it involves the image of man we entertain. We must think it anew in light of the things we can do with it or to it now and could never do before.

3. *Genetic Manipulation*

This holds even more with respect to the last object of a technology applied on man himself—the *genetic* control of future men. This is too wide a subject for the cursory treatment of these prefatory remarks, and it will have its own chapter in a later "applied part" to succeed this volume. Here I merely point to this most ambitious dream of *homo faber,* summed up in the phrase that man will take his own evolution in hand, with the aim of not just preserving the integrity of the species but of modifying it by improvements of his own design. Whether we have the right to do it, whether we are qualified for that creative role, is the most serious question that can be posed to man finding himself suddenly in possession of such fateful powers. Who will be the image-makers, by what standards, and on the basis of what knowledge? Also, the question of the moral right to experiment on future human beings must be asked. These and similar questions, which demand an answer before we embark on a journey into the unknown, show most vividly how far our powers to act are pushing us beyond the terms of all former ethics.

VIII. The "Utopian" Dynamics of Technical Progress and the Excessive Magnitude of Responsibility

The ethically relevant common feature in all the examples adduced is what I like to call the inherently "utopian" drift of our actions under the conditions of modern technology, whether it works on nonhuman or on human nature, and whether the "utopia" at the end of the road be planned or unplanned. By the kind and size of its snowballing effects, technological power propels us into goals of a type that was formerly the preserve of Utopias. To put it differently, technological power has turned what used and ought to be tentative, perhaps enlightening plays of speculative reason into competing blueprints for projects, and in choosing between them we have to choose between extremes of remote effects. The one thing we can really know of them is their extremism as such—that they concern the total condition of nature on our globe and the very kind of creatures that shall, or shall not, populate it. In consequence of the inevitably "utopian" scale of modern technology, the salutary gap between everyday and ultimate issues, between occasions for common prudence and occasions for illuminated wisdom, is steadily closing. Living now constantly in the shadow of unwanted, built-in, automatic utopianism, we are constantly confronted with issues whose positive choice requires supreme wisdom—an impossible situation for man in general, because he does not possess that wisdom, and in particular for contemporary man, because he denies the very existence of its object, namely, objective value and truth. We need wisdom most when we believe in it least.

If the new nature of our acting then calls for a new ethics of long-range responsibility, coextensive with the range of our power, it calls in the name of that very responsibility also for a new kind of humility—a humility owed, not like former humility to the smallness of our power, but to the excessive magnitude of it, which is the excess of our power to act over our power to foresee and our power to evaluate and to judge. In the face of the quasi-eschatological potentials of our technological processes, ignorance of the ultimate implications becomes itself a reason for responsible restraint—as the second best to the possession of wisdom itself.

One other aspect of the required new ethics of responsibility for and to a distant future is worth mentioning: the doubt it casts on the capacity of representative government, operating by its normal principles and procedures, to meet the new demands. For according to those principles and procedures, only *present* interests make themselves heard and felt and enforce their consideration. It is to them that public agencies are accountable, and this is the way in which concretely the respecting of rights comes about (as distinct from their abstract acknowledgment). But the *future* is not represented, it is not a force that can throw its weight into the scales. The nonexistent has no lobby, and the unborn are powerless. Thus accountability to them has no political reality behind it in present decision-making, and when they can make their complaint, then we, the culprits, will no longer be there.

This raises to an ultimate pitch the old question of the power of the wise, or the force of ideas not allied to self-interest, in the body politic. What force shall represent the future in the present? That is a question for political philosophy, and one on which I dare not voice my woefully uncertain ideas. They would be premature here anyway. For before that question of enforcement can become practical, the new ethics must find its theory, on which do's and don'ts can be based. That is: before the question of what *force,* comes the question of what *insight* or value-knowledge will represent the future in the present.

IX. The Ethical Vacuum

And here is where I come to a standstill, where we all come to a standstill. For the very same movement which put us in possession of the powers that have now to be regulated by norms—the movement of modern knowledge called science—has by a necessary complementarity eroded the foundations from which norms could be derived; it has destroyed the very idea of norm as such. Not, fortunately, the feeling for norm and even for particular norms. But this feeling becomes uncertain of itself when contradicted by alleged knowledge or at least denied all support by it. It always has a difficult time against the loud clamors of greed and fear. Now it must in addition blush before the frown or smirk of superior knowledge which has certified it as unfounded and incapable of foundation. First it was nature that was "neutralized" with respect to value, then man himself. Now we shiver in the nakedness of a nihilism in which near-omnipotence is paired with near-emptiness, greatest capacity with knowing least for what ends to use it.

It is moot whether, without restoring the category of the sacred, the category most thoroughly destroyed by the scientific enlightenment, we can have an ethics able to cope with the extreme powers which we possess today and constantly increase and are almost compelled to wield. Regarding those consequences that are imminent enough still to hit ourselves, fear can do the job—fear which is so often the best substitute for genuine virtue or wisdom. But this means fails us toward the more distant prospects, which here matter the most, especially as the beginnings seem mostly innocent in their smallness. Only awe of the sacred with its unqualified veto is independent of the computations of mundane fear and the solace of uncertainty about distant consequences. However, religion in eclipse cannot relieve ethics of its task; and while of faith it can be said that as a moving force it either is there or is not, of ethics it is true to say that it must be there.

It must be there because men act, and ethics is for the ordering of actions and for regulating the power to act. It must be there all the more, then, the greater the powers of acting that are to be regulated; and as it must fit their size, the ordering principle must also fit their kind. Thus, novel powers to act require novel ethical rules and perhaps even a new ethics.

"Thou shalt not kill" was enunciated because man has the power to kill and often the occasion and even the inclination for it—in short, because killing is actually done. It is only under the *pressure* of real habits of action, and generally of the fact that always action already takes place, without *this* having to be commanded first, that ethics as the ruling of such acting under the standard of the good or the permitted enters the stage. Such a *pressure* emanates from the novel technological powers of man, whose exercise is given with their existence. *If* they really are as novel in kind as here contended, and if by the kind of their potential consequences they really have abolished the moral neutrality which the technical commerce with matter hitherto enjoyed—then their pressure bids us to seek for new prescriptions in ethics which are competent to assume their guidance, but which first of all can hold their own theoretically against that very pressure.

In this chapter we have developed our *premises,* namely, first, that our collective technological practice constitutes a new kind of human action, and this not just because of the novelty of its methods but more so because of the unprecedented nature of some of its objects, because of the sheer magnitude of most of its enterprises, and because of the indefinitely cumulative propagation of its effects. From all three of these traits, our second premise follows: that what we are doing in this manner is, regardless of the particulars of any of its immediate purposes, no longer ethically neutral as a whole. With this exposition of the ethical question, the task of seeking an answer, and first of all a rational principle for it, only begins.

2
On Principles and Method

I. Ideal and Real Knowledge in the "Ethic of the Future"

1. *Priority of the Question of Principles*

Two questions arise as we embark on our theoretical search: What are the foundations of an ethic such as would match the new style of action? And what are the chances that its injunctions will prevail in the practical affairs of men? The first question belongs to the doctrine of moral principles, the second to the doctrine of their application—in our case, where application means public action, to political theory. The practical-political question is here all the greater as it relates to a good or a need in the distance, of which it is even harder to tell than it is of goods or needs nearby, how their knowledge among the few, assuming them to have it, can gain influence over the deeds of the many. But precisely for the sake of this influence, which is what ultimately matters, that knowledge must be safe from the suspicion of arbitrariness first of all in the minds of its purported trustees themselves—which is to say that it must not be left to the persuasion of sentiment but justified by recourse to an intelligible principle. (Or, the faith, on which knowledge of values with all its claims perhaps finally rests, must be a well-reasoned faith.) Hence the urgency of the quest for foundations: establishing them as best we can is (theoretical interest aside) of practical importance already for the sake of the authority which the prescriptions flowing from them can assert in the battle of opinions—an authority which the mere plausibility or emotional appeal of (e.g.) the proposition "We should have the future of man and the planet at heart" is insufficient to yield. For even here, the question "Why should we?" can be asked with full freedom and no frivolity, and if we leave it without an answer (even, perforce, an imperfect one) we have little right to speak of an objectively binding ethic, and can at most rely on the persuasive force of our subjective feeling. This suffices less the farther we move from the scarcely contested, and perhaps too readily granted, basic tenet that there ought to be a future inhabited by men (which hardly seems to require persuasion, although it is the most serious and momentous beginning of it all) to more specific propositions laying down that it should be a future of this sort and not of that. Toward such propositions, the question repeats itself with ever greater justification: "Why? What is the right of this particular preference? Of any preference? Indeed, of any ruling at all?" Here then, attainable truth about ultimate reasons, which is the business of philosophical knowledge, takes precedence before everything else.

2. A Predictive Science of the Long-Range Effects of Technological Action

But from the "ideal" truth about principles we must soon pass to a very different kind of truth which—being about facts—is a matter of scientific (not philosophic) knowledge: truth about predictable future conditions of mankind and the earth, on which those first, philosophic verities are to pass judgment. That judgment then will react on today's activities, from whose discerned trends those future conditions were seen, by long-range extrapolation, to follow as their certain, probable, or possible outcome. This (still theoretical) conjectural knowledge of the real and the probable in the realm of facts is thus interposed between the ideal knowledge of ethical principles and the practical knowledge of political application, which must operate with such hypothetical projections of what hope or fear have to expect—what to promote and what to prevent. We thus need a science of hypothetical prediction, a "comparative futurology," which indeed has lately begun to appear on the scene.

3. What This Science Contributes to the Knowledge of Principles: The Heuristics of Fear

It is, however, by no means the case that this connecting and concretizing link, with its depicting of future states, is really separated from the doctrine of principles: it is, rather, *heuristically* already needed within that doctrine itself. Just as we should not know about the sanctity of life if we did not know about killing, and the commandment "Thou shalt not kill" had not brought this sanctity into focus; and just as we should not know the value of truth without being aware of lies, nor of freedom without the lack of it, and so forth—so also, in our search after an ethics of responsibility for distant contingencies, it is an anticipated *distortion* of man that helps us to detect that in the normative conception of man which is to be preserved from it. And we need the *threat* to the image of man—and rather specific kinds of threat—to assure ourselves of his true image by the very recoil from these threats. As long as the danger is unknown, we do not know what to preserve and why. Knowledge of this comes, against all logic and method, from the perception of what to *avoid*. This is perceived first and teaches us, by the revulsion of feeling which acts ahead of knowledge, to apprehend the value whose antithesis so affects us. *We know the thing at stake only when we know that it is at stake.*

Because this is the way we are made: the perception of the *malum* is infinitely easier to us than the perception of the *bonum;* it is more direct, more compelling, less given to differences of opinion or taste, and, most of all, obtruding itself without our looking for it. An evil forces its perception on us by its mere presence, whereas the beneficial can be present unobtrusively and remain unperceived, unless we reflect on it (for which we must have special cause). We are not unsure about evil when it comes our way, but of the good we become sure only via the experience of its opposite. It is doubtful whether anybody would ever have praised health without at least the sight of sickness, praised sincerity without the experience of trickery, praised peace without knowing of war's misery. We know much sooner what we do not want than what we want. Therefore, moral philosophy must consult our fears prior to our wishes to learn what we really cherish.[1] And although what is most feared is not necessarily what most deserves to be feared, and still less so is its opposite the thing most deserving our desire, that is, the highest good (which, rather, may be entirely free of opposition to a defined evil)—although, in consequence, the heuristics of fear is surely not the last word in the search for goodness, it is at least an extremely useful first word and should be used to the full of its helpfulness in a sphere where so few unlooked-for words are vouchsafed us.

4. The "First Duty" of an Ethics of the Future: Visualizing the Long-Range Effects of Technological Enterprise

Now, where this word is not vouchsafed us on its own—that is, by evil already present—it becomes our duty to seek it out by an effort of reason and imagination, so that it can instill in us the fear whose guidance we need. That is the case with the "ethics of the future" which we are looking for, where that which is to be feared has never yet happened and has perhaps no analogies in past or present experience. Then the creatively imagined *malum* has to take over the role of the experienced *malum,* and this imagination does not arise on its own but must be intentionally induced. Therefore the anticipatory conjuring up of this imagination becomes itself the first, as it were introductory, duty of the ethics we are speaking about.

5. *The "Second Duty": Summoning Up a Feeling Appropriate to What Has Been Visualized*

But one sees at once that this imagined and distant evil, which is not mine, will not arouse fear as naturally and spontaneously as does a present danger which threatens myself or those near to me. As little, therefore, as the *image* (idea) of the faraway fearsome comes about all by itself, does the *fear* of it do so: it, too, must be procured with an assist from ourselves, namely, a prior sensitizing of our feeling to this kind of stimulus. Responsiveness to it has to be acquired on purpose. The psychology of the matter is thus not as simple as it was for Hobbes, who also, instead of love for a *summum bonum,* made fear of a *summum malum,* namely, the fear of violent death, the starting point of morality. This is an evil well known or imagined, its potential ever present, its acute threat arousing the extreme of fear as the most compulsive reaction of our innate instinct of self-preservation. The imagined fate of future men, let alone that of the planet, which affects neither me nor anyone else still connected with me by the bonds of love or just of coexistence, does not of itself have this influence upon our feeling. And yet it "ought" to have it—that is, *we* should allow it this influence by purposely making room for it in our disposition. The fear in question then cannot be, as in Hobbes, of the "pathological" sort (to use Kant's term), which compulsively overcomes us in the face of its object, but rather a spiritual sort of fear which is, in a sense, the work of our own deliberate attitude. Such an attitude must be cultivated; we must educate our soul to a willingness to *let* itself be affected by the mere thought of possible fortunes and calamities of future generations, so that the projections of futurology will not remain mere food for idle curiosity or equally idle pessimism. Therefore, bringing ourselves to this emotional readiness, developing an attitude open to the stirrings of fear in the face of merely conjectural and distant forecasts concerning man's destiny—a new kind of *éducation sentimentale*—is the second, preliminary duty of the ethic we are seeking, subsequent to the first duty to bring about that mere thought itself. Informed by this thinking, we are obliged to lay ourselves open to the appropriate fear.

It is obvious that both duties presuppose a more basic ethical principle from which their binding force derives, and which must be made explicit before one can see that they have this force, that is, that they are indeed duties. Of this, we shall hear soon.

6. *The Uncertainty of Prognostications*

Let us first turn once more to what we claimed to be the specific duty of *thought* laid upon us by the present state of things. We said that the truth it was to seek, namely, about things to come, was a matter for scientific knowledge. This is little more than a truism. For just as it is only through science that those enterprises are made possible whose later consequences we are told to discover by extrapolation, so this extrapolation in turn clearly demands at least the same degree of science as is embodied in those technological deeds themselves. In fact, however, it demands a still higher degree. For the degree which suffices for the short-range prediction intrinsic to each work of technology by itself—the engineer's prediction of its working—is on principle inadequate for the long-range prediction of the combined working of all of them: and at this, the ethically required extrapolation must aim. The certainty which the one enjoys, and without which the whole technological enterprise could not function at all, is forever denied to the other. Here we need not elaborate the reasons for this. We only name the immense complexity of the societal and biospheric totality of effects, which defies all calculation (supercomputers included); the essentially unfathomable nature of man, which always springs surprises on us; and the impossibility to predict future inventions, which would amount to preinventing them. More about this later. In any case, the required extrapolation demands an exponentially higher degree of science than is already present in the technology from which it is to be extrapolated. And since this represents in each case the apex of existing science, the knowledge called for is of necessity always "not yet" available, which means: to foreknowledge never, and at best (if at all) to hindsight.

7. Knowledge of the Possible: Heuristically Sufficient for the Doctrine of Principles

This, however, does not preclude the projection of probable or arguably possible end effects; and the mere knowledge of *possibilities,* though certainly insufficient for cogent prediction, is fully adequate for the purposes of a *heuristic* casuistry that is to help in the spotting of ethical principles. Its means are thought experiments, which are not only hypothetical in the assumption of premises ("*If* this is done, then that follows") but also conjectural in the inference from "if" to "then" (" . . . then such a thing *can* follow"). It is the content, not the certainty, of the "then" thus offered to the imagination as possible, which can bring to light, for the first time, principles of morality heretofore unknown for lack of the actual occasions to which they could apply and which would have called attention to them. The perceived possibility can now take the place of the actual occasion; and reflection on the possible, fully unfolded in the imagination, gives access to new moral truth. But *this* truth belongs to the sphere of ideas, that is, it is just as much a matter of *philosophical* knowledge as is the truth of that grounding first principle we have yet to supply. Accordingly, *its* certainty is not dependent upon the degree of certainty of the factual, scientific projections which provided paradigmatic material for it. Whatever the ultimate accreditation for this kind of truth—be it the self-evidence of reason, an *a priori* of faith, or a metaphysical decision of the will—its pronouncements are apoaictic, whereas those of the hypothetical thought experiments can at best be probabilistic. This is enough where they are meant to serve not as proofs but as illustrations. What is here contemplated, therefore, is a casuistry of the imagination which, unlike the customary casuistries of law and morality that serve the trying out of principles already known, assists in the tracking and discovering of principles still unknown. The serious side of science fiction lies precisely in its performing such well-informed thought experiments, whose vivid imaginary results may assume the heuristic function here proposed. (See, e.g., A. Huxley's *Brave New World*.)

8. *Knowledge of the Possible: Seemingly Useless for Applying the Principles to Politics*

Thus the heuristics of which we talk is a prognostic which extrapolates from presently recognizable trends in the technologic-industrial process. The fact that such extrapolations are no more than conjectural, and each can be opposed plausibly by others without conclusive estimation of the relative probabilities, we found to be innocuous for the doctrine of principles, which needs no more than demonstrated possibilities. But it must be admitted now that this same uncertainty of all long-term projections becomes a grievous weakness when they have to serve as prognoses by which to mold behavior—that is, in the practical-political *application* of whatever principles were apprehended with the help of the heuristic casuistry. For there, the envisaged distant outcome should lead its beholder back to a decision on what to do or abstain from now: and one demands, not unreasonably, a considerable certainty of prediction when asked to renounce a desired and certain near-effect because of an alleged distant effect, which anyway will no longer touch ourselves. To be sure, in the truly capital issues of ultimate destiny, the order of magnitude of the unwilled long-term effects so far exceeds that of the intended short-term effect that it ought to outweigh quite some disparity in certainty. Nevertheless, the qualification "merely possible" that mars the projections and is inseparable from the theoretical weakness of all the extrapolation methods at our disposal, is easily fatal: for, naturally, it means that something else is also possible—and who is to say it is not "just as possible"? Then, interest or inclination or opinion can each time select, from among the possible prognoses, the one most favorable to the project anyway preferred by it and best suited to the clamor of the hour; or brush them all aside with the agnostic decree that we generally know too little to sacrifice the known for the unknown; and for the rest argue that there will still be time for corrections "en route," when "we" (meaning those after us) will see what happens. Being so much in the dark, why not trust our luck including that of posterity? But in this way, all the gains of our hypothetical heuristics are kept from timely application by the inconclusiveness of the prognostics, and the finest principles must lie fallow until it is, perhaps, too late.

II. Prevalence of the Bad over the Good Prognosis

But just this uncertainty, which threatens to make the ethical insight ineffectual for the long-range responsibility toward the future—an uncertainty not confined, of course, to the prophecy of doom—has itself to be included in the ethical theory and become the cause of a new principle, which on its part can yield a not uncertain rule for decision-making. It is the rule, stated primitively, that the prophecy of doom is to be given greater heed than the prophecy of bliss. The reasons for this shall be briefly pointed out.

1. *The Probabilities in the Great Wagers*

First, the mere probability relation between lucky and unlucky results of unknown experiments is generally like that between either hitting or missing a target. The hit is only one of innumerable alternatives which otherwise are more or less wide misses. And although in small matters we may allow many such misses in order to benefit from the rarer chances of success, we may allow but few where greater things are concerned. And in the really great, irreversible ones, which go to the roots of the whole human enterprise, we really must allow none. Natural evolution works with small things, never plays for the whole stake, and therefore can afford innumerable "mistakes" in its single moves, from which its patient, slow process chooses the few, equally small "hits." The big enterprise of modern technology, neither patient nor slow, compresses— as a whole and in many of its single projects—the many infinitesimal steps of natural evolution into a few colossal ones and forgoes by that procedure the vital advantages of nature's "playing safe."

To the causal *extent* is thus added the causal *tempo* of technological interference with life's system. It is therefore far from true that "taking his own evolution in hand"—that is, replacing blind and slow-working chance with conscious and fast-working planning—would give man a surer prospect of evolutionary success. On the contrary, it would inject entirely new elements of insecurity and hazard, which rise in proportion as the stakes are raised and the times are shortened in which the inevitable (and no longer small) mistakes can be corrected. And inevitable they are, not just because of human fallibility in general, but because that which is "short" for natural evolution is still very long for man, and in forcing this time contraction on evolutionary change he, by the same token, forces a vastly lengthened radius on his own actions—thus infesting them with that essential inadequacy of his long-range predictive powers that we discussed before. If we add to this the disparity of hit-or-miss chances that we found to prevail here, we must bow to the command to allow, in matters of such capital eventualities, more weight to threat than to promise and to avoid apocalyptic prospects even at the price of thereby perhaps missing eschatological fulfillments. It is the command of caution in the face of the revolutionary style which the evolutionary "either-or" technique assumes under the aegis of technology with its inherent playing for the whole stake—so foreign to natural evolution.

2. *The Cumulative Dynamics of Technological Developments*

Second, beyond these general observations, we must note that there is something ominously peculiar about the "en route" which after all—so one might contend—will still be there for us (and later ones) to make the necessary corrections as we go along. Experience has taught us that developments set in motion by technological acts with short-term aims tend to make themselves independent, that is, to gather their own compulsive dynamics, an automotive momentum, by which they become not only, as pointed out, irreversible but also forward-pushing and thus overtake the wishes and plans of the initiators. The motion once begun takes the law of action out of our hands, and the accomplished facts, created by the beginning, become cumulatively the law of its continuation. Granted then that we can "take our own evolution in hand," it will slip from that hand by the very impulse it has received from it; and here more than anywhere else applies the adage that we are free at the first step but slaves at the second and all further ones. So we have to add to the first observation—that the speed of technologically fed developments does not leave itself the time for self-correction—the further observation that in whatever time is left the corrections will become more and more difficult and the freedom to make them more and more restricted. This heightens the duty to that vigilance over the beginnings which grants priority to well-grounded possibilities of disaster (different from mere fearful fantasies) over hopes, even if no less well grounded.

3. *The Sacrosanctity of the Subject of Evolution*

Third, on a less pragmatic plane, there is the heritage of a past evolution for us to preserve—which heritage cannot be all bad in our case, if only because it has bequeathed to its present incumbents the (self-proclaimed) capacity to be judges of what is good and what is bad. This heritage, however, can perish. In a generally miserable condition, where change itself can seem to promise improvement, one can (reasoning on the lines of "the proletarian has nothing to lose but his chains") blithely wager one's all in the conviction that any winnings in the game can only be something better, and even losing the whole stake will not be much of a loss. But the advocates of the utopian wager cannot appeal to this logic. For their enterprise is replete with pride in their knowledge and in their knowing capacity, for the gift of which they have to thank, after all, past natural evolution. They thus either disparage this evolution with a readiness to overthrow its outcome because it is lacking, but then have disqualified themselves (being that outcome) for the task of improving it; or they assert that very qualification, but then have affirmed its premise, their inherited natural constitution, as what precisely is not to be tampered with.[2]

There is, however, a third alternative, namely, to refrain from both disparagement and claim to superior qualification alike, and to simply say: Since nothing is sanctioned by nature and therefore everything is permitted to us, we have full freedom for creative play that is guided by nothing but the whim of the playing impulse and makes no claim other than to master the rules of the game, that is, the claim of technical competence. This standpoint of nihilistic freedom, self-exempted from the need for justification, is indeed free of internal contradiction, but we need not discuss it, since we will certainly not entrust our fate to professed irresponsibility. Some kind of authority must be asserted for the determination of models, and unless we subscribe to dualism and say that the cognitive subject is from above the world, this authority can only base itself on an essential sufficiency of our nature such as it has evolved within this world. Now, this innate sufficiency of human nature, which we must posit as the enabling premise for any creative steering of destiny, and which is nothing other than the sufficiency (albeit fallible) for truth, valuation, and freedom, is a thing unique and stupendous to behold in the stream of becoming, out of which it emerged, which in essence it transcends, but by which it can also be swallowed again. Its possession therefore, as much as we were granted of it, purports that there is something *infinite* for us to preserve in the flux, but something infinite also to lose. Most evidently, the authority which it imparts can never include the disfiguring, endangering, or refashioning of itself. No gain is worth this price, no hope of gain justifies this risk. And yet, today, this very share of transcendence too is in danger of being thrown into the crucible of biotechnological alchemy—as if the enabling condition for all our freedom to revise the given were itself among the revisables. The logical folly of it aside, the ingratitude it implies for the inheritance badly accords with the extreme use of its gift which the very attempt at its revision represents. About gratitude, piety, and respect as ingredients of an ethic called upon to stand guard over the future in the technological tempest of the present, and which cannot do so without the past, we will have more to say later. For now, the important thing is only to note that, among the stakes risked in the game, there is one of metaphysical rank (physical as its origins may be), an "absolute" which, as a supreme and vulnerable trust, lays upon us the supreme duty to preserve it intact. This duty is beyond comparison superior to all the injunctions and wishes of a meliorism in the peripheral

zones, and, where it is concerned, the question is no longer one of weighing chances of finite profit and loss but one of contraposing the risk of
infinite loss against chances of finite gains. No weighing (e.g., of probability differentials) has still a place between these incommensurables. Thus
when it comes to this core phenomenon of our humanity, which is to be
preserved in its integrity at all costs, and which has not to await its
perfection from the future because it is already whole in its essence as
we possess it—when it comes to this, then indeed the well-enough founded
prognosis of doom has greater force than any concurrent prognosis of
bliss, be it equally well or even better founded, but which of necessity
pertains to matters on a lower plane. The reproach of "pessimism" leveled
at such partisanship for the "prophecy of doom" can be countered with
the remark that the greater pessimism is on the side of those who consider
the given to be so bad or worthless that every gamble for its possible
improvement is defensible.

III. The Element of Wager in Human Action

So much for the reasons in support of the "prophecy rule." Let us
formulate now the ethical principle behind it, from which those reasons
also draw their force. We started out with the observation that the *uncertainty* of all long-term projections, which by the equipoise of alternatives threatens to paralyze the application of principles to the sphere
of facts, is in its turn to be taken as a fact, for the right treatment of which
ethics has to find a principle which itself is no longer an uncertain one.
What I have discussed so far was not this principle itself but the practical
rule in which it issues, namely, to give in matters of a certain magnitude—
those with apocalyptic potential—greater weight to the prognosis of doom
than to that of bliss. The premise of the whole argument was that both
today and in the future we will have to deal with actions of just that
magnitude, which is itself a *novum* in human affairs. This *novum* renders
obsolete the tacit standpoint of all earlier ethics that, given the impossibility of long-term calculation, one should consider what is close at hand
only and let the distant future take care of itself. This still goes for the
private sphere of action where enticing as well as threatening distant
perspectives are nothing more than idle fantasies which should be allowed
neither pragmatic nor moral influence upon short-term decisions. To disregard them, that is, to disregard vain hopes and fears, is here the only
rule appropriate to the uncertainty; and not to brood over the unknown
is a presupposition of effective virtue. But in the new action-sphere of
planetary technological planning, the fantasies are not idle: projection into
the distance belongs to its nature and duty, and *its* uncertainty must
therefore be met by another rule.

We know its contents. Its principle we shall learn when we reflect on the element of *game* or *bet* contained in all human action concerning its issue and side effects, and ask ourselves for what stake it is ethically permissible to gamble.

1. *May I Stake the Interests of Others in My Wager?*

A first answer that suggests itself would be that, strictly speaking, one may wager nothing which one does not own (leaving open whether one may wager everything one owns). But we cannot live with this answer, because in the inextricable interweaving of human affairs as of all things, it is nowise avoidable that my action will affect the fate of others; and thus any staking of my very own is always a staking also of something that belongs to others and to which I have properly no right. This element of guilt must be shouldered in all action (of which quietistic inaction would only be another variety). This is true not only for the faceless guilt which I shall never know and just must summarily assume in its inevitability, but also for the determinate guilt I can know and foresee. "He who acts," says Goethe, "is always unscrupulous" (*Der Handelnde ist immer gewissenlos*), and by this he presumably means the readiness to contract this guilt. How much of such "unscrupulousness" the higher moral conscience can allow—i.e., how far we may go in knowingly violating, or just putting at risk (as "stakes"), the interests of others in our projects—to determine this in each instance is a matter for the casuistics of responsibility and can in general not be laid down *a priori* in the doctrine of principles. Only wantonness and levity are inadmissible on principle—when wagering what is others' and even what is mine; that is to say, the "unscrupulousness" must not be thoughtless. And wanton it would be, for instance, to stake things of importance in the bid for vain goals. Even there, frivolity or foolhardiness regarding one's own well-being and even life is a right not really to be disputed: at most one can say of it that it is limited by an opposing duty, but not that it is thereby annulled. Only the inclusion of others in my "wager" makes frivolity unacceptable. With this stipulation observed, some such inclusion must be allowed if there is to be any action at all.

2. *May I Put All the Interests of Others at Stake?*

Perhaps, then, we should qualify our first, exclusionary answer, which we had to reject as too sweeping, by adding that it must never be the *totality* of the interests of others, above all not their lives, which I wager in my game of uncertainty. And indeed, in the pursuit of narrowly selfish interests, this prohibition has unconditional force, if only because of the disproportion which would normally exist between the partial nature of the interests pursued (my own) and the total nature of the interests risked (others'); in fact, however, it holds even where not my advantage merely but my very life is at stake. But does it also hold for the pursuit of selfless goals? Especially such that are pursued in the interest of those "others" put at risk themselves?

One would not deny the statesman the right to risk his nation's existence for its future if really ultimate issues are at stake. In this manner the awesome but morally defensible decisions about war and peace come about when, for the future's sake, the stake is the future itself. It has to be added that this should never happen because of the enticement of a wonderful future but only under the threat of a terrible future; not to gain a supreme good (an immodesty in any case) but only to prevent a supreme evil. The latter consideration always has priority and alone has the excuse of dire necessity, for one can live without the supreme good but not with the supreme evil. There never is a good reason for the alternative of win all or lose all. But seeking to save that without which all else is deemed worthless, at the peril of losing all in the attempt, may be morally justified and even commanded. Thus, minding this asymmetry between goods desired and evils shunned, even the proposition that never the whole of the interests of others must be staked in the wager of action is not unconditionally valid.

3. *Meliorism Does Not Justify Total Stakes*

However, this proviso—namely, that only the prevention of greatest evil and not the ushering in of the greatest good can justify in certain situations the total risk of others' interests for their own good—excludes from its qualified license the great risks of technology. For these are not undertaken to preserve what exists or to alleviate what is unbearable, but rather to continually improve what has already been achieved, in other words, for *progress,* which at its most ambitious aims at bringing about an earthly paradise. It and its works stand therefore under the aegis of arrogance rather than of necessity; declining some of its beckonings will hit the optional and marginal beyond the necessary, while pursuing them may injure, in the end, the inner core. Thus, here where the protection of the proviso does not reach, the proposition that I must never with my action put at risk "the *whole*" interest of the others involved (who in this case are future persons) again acquires force.

4. *Mankind Has No Right to Suicide*

As the capstone of our argument, let us now add that the sum total of the interests at stake in the "bet" of technological progress has an incomparably wider compass than anything which normally is at stake in human decisions. Even when in the fateful hour the political leader hazards the whole existence of his tribe, his city, his nation, he yet knows that even should they be destroyed, mankind and a living world on earth will go on. Only in the framework of this overarching supposition is the single hazard, in certain extreme cases, morally defensible. But not even for saving his nation's life must the statesman use means that can destroy mankind. Now, among the possible works of technology there are some which cumulatively have just that global extent and depth, namely, the possibility to endanger the whole existence or whole being of man in the future. The statesman, in making his fateful decision, can ideally assume the consent of those for whom, as their agent, he is making it. No consent to their nonexistence or dehumanization is obtainable from the humanity of the future, nor can it be assumed; and were it nevertheless imputed to them (an almost insane imputation), it would have to be rejected. For there is (as has yet to be shown) an *unconditional duty* for mankind to exist, and it must not be confounded with the conditional duty of each and every man to exist. The right of the individual to commit suicide is morally arguable and must at least for particular circumstances be conceded: under no circumstances has mankind that right.

5. *The Existence of "Man" Must Never Be Put at Stake*

Herewith we have at last found a *principle* that forbids certain technologically feasible "experiments," and of which that rule for decision making—to give the bad prognosis precedence over the good—was the pragmatic expression stated in advance. The ethical axiom which validates the rule is therefore as follows: Never must the existence or the essence of man as a whole be made a stake in the hazards of action. It follows directly that bare possibilities of the designated order are to be regarded as unacceptable risks which no opposing possibilities can render more acceptable. The rule that even imperfect palliatives are to be preferred to a promising radical cure from which the patient may die is valid for the life of mankind though not always for the individual patient.

We have here an inversion of Descartes's principle of doubt. In order to ascertain the indubitable truth we should, according to Descartes, equate everything doubtful with the demonstrably false. Here on the contrary we are told to treat, for the purposes of decision, the doubtful but possible as if it were certain, when it is of a certain kind. It is also a subspecies of the Pascalian wager without the selfish-eudaemonistic and ultimately unethical quality thereof. According to Pascal, pure calculation in the wager between the brief and questionable pleasures of this world and the *possibility* of eternal bliss or damnation in the next, demands that one bank on just this extreme possibility, since a comparison of the chances of profit and loss on both sides reveals that, in choosing the second alternative, even if its object, eternal life, does not exist, one will have lost but a small thing in the temporal life; while if it does exist, something infinite will have been gained and an infinite loss avoided. Contrariwise, the choice in favor of temporal life could at best (i.e., if there is no such thing as eternal life) mean a small gain, but in the opposite case (with eternal damnation) would come out an infinite loss. This "go-for-broke" calculation of risks, objectionable also in other respects, is in error already by the fact that, in proportion to the *nothing* which is here accepted among the risks, every *something* (including the things of fleeting temporal existence) is of infinite magnitude; and thus the second alternative (betting all upon a possible eternity while sacrificing the given temporality) *also* bears the risk of *infinite* loss. There must be more than mere possibility: a *faith* must assert that an eternity awaits us—and then the option is no longer purely a bet. Absolute uncertainty, however, can on principle not count in any computation against the relative certainties of the given. *Our* ethical principle of the wager is not open to this objection. For it forbids us precisely to incur the risk of nothingness, that is, to allow the presence of its possibility among the chances of our choice. It forbids, in short, any *va-banque* game in the affairs of humanity. Nor does it pit what is essentially unknowable and even beyond imagination against the knowable or imaginable objects of choice, but rather sets the totally *unacceptable* over against the more or less acceptable *within* the imaginable finite itself. Above all, it is morally commanding, not just a calculation of advantages presented to self-interest; and it commands on the basis of a primary duty to opt for being and against nothingness.

This principle for the treatment of uncertainty is itself not uncertain at all, and it binds us unconditionally—that is, not just as an advice of moral prudence but as an unqualified command—*provided* we accept the responsibility for what will be. Under such responsibility, *caution,* otherwise a peripheral matter of our discretion, becomes the core of moral action. Now, that presupposition for it just named, that *we are* indeed "responsible," even that there *is* such a thing at all, was throughout our preceding discourse tacitly assumed, but nowhere proved. The *principle* of responsibility as such—the starting point of ethics—has not yet been demonstrated. To this task, for which formerly one might well have invoked the aid of heaven which it so direly needs—and more so today when it no longer can benefit from even gazing in that direction—we shall now turn.

IV. The Duty to Ensure a Future

1. *The Nonreciprocity of Duties to the Future*

At the outset it must be realized that what we require of our principle is not supplied by the traditional idea of rights and duties—the idea grounded upon reciprocity, according to which my duty is the counterpart of another's right, which in turn is seen as the like of my own right: once certain rights of another are established, then my corresponding duty to respect them and (adding the further idea of a positive responsibility) where possible to further them is also established. This scheme fails for our purposes. For only that *has* a claim that *makes* claims—for which it must first of all *exist*. Everything alive makes a claim to life, and perhaps this is a right to be respected. The nonexistent makes no demands and can therefore not suffer violation of its rights. It may have rights when it exists, but it does not have them by virtue of the mere possibility that it will one day exist. Above all, it has no right to exist at all before it in fact exists. The claim to existence begins only with existence. But the ethic we seek is concerned with just this not-yet-existent; and *its* principle of responsibility must be independent of any idea of a right and therefore also of a reciprocity—so that within its framework the question jokingly invented for the situation: "What has the future ever done for me? Does it respect my rights?" cannot possibly be asked.

2. *The Duty Regarding Posterity*

Now, we are familiar already in traditional morality with *one* most widespread case (deeply moving even to the spectator) of elementary *nonreciprocal* responsibility and duty, which moreover enjoys immediate recognition and spontaneous enactment: the duty to the children one has brought forth and who without the continuance of procreation into provision and care must perish. To be sure, one may expect some return from them, in old age, for the love and effort expended upon them, but this is certainly not the precondition or motive for doing it, and still less a condition of the responsibility itself one owes toward the child—which, on the contrary, is unconditional. This is the only class of fully selfless behavior supplied by *nature;* and indeed, it is in this one-way relationship to dependent *progeny,* given with the biological facts of procreation, and *not* in the mutual relationship between independent adults (from which, rather, springs the idea of reciprocal rights and duties) that one should look for the origin of the idea of (basically one-sided) responsibility in general; and its constantly demanding sphere of action is the original site of its practice. Without this fact and the sex relation bound up with it, we could understand neither the arising of farsighted providence nor of selfless care among rational beings, be they ever so social. (We will later make use of this observation, which to my knowledge has never been sufficiently appreciated in moral philosophy.) Here is the archetype of all responsible action, which fortunately requires no deduction from a principle, because it is powerfully implanted in us by nature or at least in the childbearing part of humanity.

If, nevertheless, as ethical theory requires, we reflect on the ethical principle operative here (and males must perhaps sometimes be reminded of it), then we see that the duty toward children and the duty toward later generations are not the same. The duty to care for the now-existing child produced by us can, even without the incentive of feeling, readily be based upon the responsibility of the *cause* for its effect, that is, upon our having *originated* his existence, and then, in consequence, upon the *right* that henceforth pertains to this existent and issues from itself. Thus, despite nonreciprocity, the obligation here rests on the classical principle of rights and duties, albeit both one-sided in this case (the suckling has no complementary duty, the parent no complementary right). But something other than the duty arising *from* procreation, which already faces a right arising from existence, would be the duty *to* such procreation, to the generation of children, to reproduction in general. This duty, if it exists, is far harder to prove and, if at all, then surely not from the same principle; and a *right* of the unborn to being born (more precisely: of the ungenerated to generation) is simply not arguable from any principle whatsoever. Thus we would here deal with a duty that is not the counterpart of another's right—unless it be the right of God the Creator over against his creatures to whom, with the bestowal of existence, he has entrusted this continuation of his work.

3. *The Duty toward the Existence and the Condition of*
Future Generations

It is this sort of duty that is involved in a responsibility for future mankind. It charges us, in the first place, with ensuring that there *be* a future mankind—even if no descendents of ours are among them—and second, with a duty toward their *condition,* the quality of their life. The first duty contains within itself the duty of procreation (even if not necessarily that of everybody) and like the latter is not simply derivable, by extension, from the duty of the begetter toward the life already begotten. If there is such a duty, as we would like to suppose, our discourse has yet to show a sanctioning ground for it.

a) Does the duty to posterity need proof?   Now it is arguable that we can save ourselves the question of whether such a duty exists, and with it the thorny task of discovering its sanctioning ground, since we need have no fear for the permanence of the procreative drive; and as to possible external causes of extinction due to ourselves (e.g., truly lethal poisoning of the environment) one may hold that only the most unlikely combination of the most unlikely and colossal idiocies on our part could bring them about—which, with all due respect for the dimensions of human stupidity or folly, we need not consider as a serious possibility. Should we then not just assume continued existence as such and directly turn to the more concrete subject of our duties toward its future quality? These have the advantage of being more nearly derivable from familiar principles of morality, and their observance anyway helps also secure the mere existence of mankind which they presuppose.

Both points are correct. Surely, the dangers which threaten the quality of future life are in general the same as those which, magnified, threaten survival itself, and avoiding the ones is *a fortiori* avoiding the others. And a possible derivation of duties toward the condition (rather than the existence *per se*) of future mankind from the familiar idea of rights and duties might go somewhat like this: Since there will in any case be later men, who had no say in their coming to exist, they will in their time have the right to accuse us who came before them of being the originators of their misfortune—if we have spoiled for them the world or the human constitution through careless and avoidable deeds. Whereas for their existence they can hold responsible only their immediate begetters (and even there have a right to complain only if the parents' right to progeny can be questioned on specific grounds), for the *conditions* of their existence they can hold distant ancestors or, more generally, the originators of these conditions responsible. Thus, from the *right* of the anticipated being of later subjects, there issues for us today, as the causative agents, an answering *duty,* which makes us responsible to them with those of our deeds whose effective range extends into such dimensions of time, space, and depth.

b) The priority of the duty to existence   But as correct and perhaps pragmatically sufficient as this line of reasoning is, it is insufficient for ethical theory. For, first of all, the conscientious pessimist could, if the prospect is gloomy enough, declare those to be irresponsible who carry on the business of reproduction "notwithstanding" and, for his own part, decline responsibility for the fruits of a folly from which he has abstained. In other words, on the view that after all there do not have to be humans at all costs (and we have not yet ruled out this view), he can make the desirability or mandatoriness of a future humanity depend upon the foreseeable conditions of its existence, instead of conversely allowing the unconditional mandatoriness of such an existence to dictate the conditions we ought to provide for it. (This is an extension of the argument which I often heard from despairing emigrant couples in the Hitler period—that one must not "bring children into such a world.")

Second, there is the more decisive consideration that the anticipated accusation by our future victims would arise from a supposed grievance over their lot and thus would fail to arise if they were in agreement with it, or even felt quite happy with their condition. But such an agreement or happiness might be the last thing we should wish for a future humanity, namely, if it were purchased with the dignity and vocation of man. Thus, it could be that we would rather have to accuse ourselves of the fact that *no* accusation against us issues hence. The absence of protest would then itself be the gravest accusation; but the accuser in that case would *not* be the future injured party, but rather we ourselves. What does this mean?

It means that in the final analysis we consult not our successors' *wishes* (which can be of our own making) but rather the "ought" that stands above both of us. To make it impossible for them to be what they *ought* to be is the true crime, behind which all frustration of their desires, culpable as it may be, takes second place. This means, in turn, that it is less the *right* of future men (namely, their right to happiness, which, given the uncertain concept of "happiness," would be a precarious criterion anyway) than their *duty* over which we have to watch, namely, their duty to be truly human: thus over their *capacity* for this duty—the capacity to even attribute it to themselves at all—which *we* could possibly rob them of with the alchemy of our "utopian" technology. To stand guard over this onerous endowment of theirs is *our* cardinal duty toward the future of humanity as such (its existence presupposed), from which all specified duties toward the well-being of future humans are then derivative. These various, concrete duties may well be subsumable under the known ethic of solidarity, of sympathy, of equitableness, indeed, even of compassion, by whose standards, and through a transference from our own hopes and fears, joys and sorrows, we accord to those future individuals, in a kind of fictive contemporaneity, the same right which this ethic also accords to those living now, and which it enjoins us to respect: Our respecting it "prospectively" in its future subjects as well becomes our specially emphatic responsibility through the totally one-sided causality of our role as *originators* of their condition. Here, then (as said before), our duty still answers to a *right* "existing," that is, anticipated as existing, on the others' side: their right, when their time comes, to a worthwhile quality of life, both as to its internal powers and its external conditions. But this duty toward the *quality* of future life, provided it comes to it, stands itself under the condition of the aforementioned duty to ensure that there *be* future life and with it future subjects of possible rights—and *this* duty answers to no right whatsoever but rather gives *us* in the first place the (otherwise dubious) right to bring beings like ourselves into the world without their choice. The individual right (to procreation) here follows from the general duty to continued human existence, and not vice versa. And while the exercise of this right then entails particular duties toward those brought thereby into existence—duties with whose principle we are well acquainted—such duties along with their principle are as a whole subordinate to that primary duty which, totally one-sided, empowers us vis-à-vis all those after us, not so much to make them the gift of their existence (which ill agrees with forcing it upon them) as to *tax* them with it—namely, with the very sort of existence that is capable of the burden which is the true object of our duty to bestow that existence on them. Whether they also *wish* for this burden we would not ask them even if we could. That there *be* future bearers of it, so as to ensure *its* very perpetuity in the world, comes first in the hierarchy of duties. But of course, to impose it on future subjects presupposes that we have not prejudiced their capacity to bear it. This, then, is the first, qualitative duty toward the *condition* of our successors: itself already derivative from the duty to make them *exist,* it in turn presides over all the other, various duties toward them, for instance those relating to their chances of happiness.

c) *The first imperative: that there be a mankind* It is therefore not the case (as first suggested) that we can leave aside the question of our responsibility for there being a future mankind and simply turn to our duties toward the mankind that will in fact be there—that is, to providing for its quality and condition. On the contrary, the first rule for the mandatory quality can solely be gained from the imperative of existence, and all further rules are subject to its criterion, which no eudaemonistic ethic alone nor any ethic of compassion can supply. Much is admissible under either of these which that imperative forbids, and much may be protested by them which it commands. The first rule is, therefore, that no condition of future descendants of humankind should be permitted to arise which contradicts the reason why the existence of mankind is mandatory at all. The imperative that there be a mankind is the first one, as far as man alone is concerned.

4. *Ontological Responsibility for the Idea of Man*

With this imperative we are, strictly speaking, not responsible to the future human individuals but to the *idea* of Man, which is such that it demands the presence of its embodiment in the world. It is, in other words, an ontological idea, which does not, however, as the "ontological proof" alleges concerning the concept of God, guarantee the existence of its subject already with the essence—far from it!—but says that such a presence *ought* to be and to be watched over, thus making it a duty to us who can endanger it. It is this ontological imperative, emanating from the idea of Man, that stands behind the prohibition of a *va-banque* gamble with mankind. Only the idea of Man, by telling us *why* there should be men, tells us also *how* they should be.

5. *The Ontological Idea Generates a Categorical, Not a Hypothetical, Imperative*

The Kantian distinction between the hypothetical and categorical imperative, intended for the ethic of contemporaneity, applies as well to the ethic of responsibility for the future which we are here trying to define. The hypothetical (of which there are many) says: *If* there are human beings in the future—which depends on our procreation—then such and such duties are to be observed by us toward them in advance. But the categorical commands only *that* there be *human beings,* with the accent equally on the *that* and the *what* of obligatory existence. For me, I admit, this imperative is the only one which really fits the Kantian sense of the categorical, that is, the unconditional. Since *its* principle is not, as with Kant, the self-consistency of reason giving to itself laws of conduct, that is, not an idea of *action* (of which some kind or other is presupposed to occur anyway), but is rather the idea of possible agents in general, for whom it claims that such *ought to exist* (fallible as they must be), and is thus "ontological," that is, an idea of *being*—it follows that the first principle of an "ethic of futurity" does not itself lie *within* ethics as a doctrine of action (within which thereafter all duties toward future beings belong), but within *metaphysics* as a doctrine of being, of which the idea of Man is a part.

6. *Two Dogmas: "There Is No Metaphysical Truth" and "No Path Leads from 'Is' to 'Ought' "*

The last contention contradicts the most firmly entrenched dogmas of our time: that there is no metaphysical truth, and that no "ought" can be derived from "being." The latter has never been seriously examined and is true only for a concept of being that has been suitably neutralized beforehand (as "value free")—so that the nonderivability of an "ought" from it follows tautologically. To expand this trivial conclusion into a general axiom is equivalent to asserting that no other concept of being is possible, or: that the one serving as the premise here (ultimately borrowed from the natural sciences) is the true and complete concept of being. Thus, with the very assumption of such a concept of being, the rigid separation of "is" and "ought" reflects in itself already a definite *metaphysics,* which can only boast the critical (Occamistic) advantage that it makes the most parsimonious assumption about being (therewith, of course, also the most meager for the purpose of explaining the phenomena, hence at the price of the latters' impoverishment).

But if the dogma that no path leads from "is" to "ought" is, by reason of its ontological presupposition, a metaphysical proposition, then it falls under the interdict of the first and more fundamental dogma: that there is no metaphysical truth. *This* proposition has its own presupposition to which *its* validity is tied. Just as the dogma of "is and ought" presupposes a definite concept of being, so does the denial of metaphysical truth presuppose a definite concept of knowledge, for which it is indeed true: "scientific" truth is not to be had about metaphysical objects—once again a tautological conclusion since science is just concerned with physical objects. So long as it is not indisputably shown that this exhausts the whole concept of knowledge, the last word on the possibility of metaphysics has not yet been spoken. But be this as it may, even its conceded denial would not be an objection peculiar to the ethic we are seeking. For in every other ethic as well, in the most utilitarian, most eudaemonistic, most this-worldly, etc., a tacit metaphysics is imbedded ("materialism," e.g., would be one), and they are not a whit better off in this regard. What is special about our case is only that the inherent metaphysics cannot remain hidden but must come to light—which for the proximate business of mere ethics is a tactical disadvantage, but for the cause of truth rather more of an advantage in the end. It is the advantage of compelling us to give an account of the metaphysical *grounds* of obligation. For although the negative thesis regarding "is and ought" likewise implies a metaphysical thesis, its proponent simply can acquiesce in the universally shared ignorance in matters metaphysical and retreat to the then valid methodological superiority of the minimal assumption, that is, of the negation over the affirmation. The affirmation (that "is" can yield "ought") is denied this asylum and its advocate must bring forward, if not a proof, at least a reasonable ontological argument for his more immodest assumption. Thus, for him the metaphysical attempt is necessary, while the "minimalist," invoking Occam, can spare himself the effort.

7. *Concerning the Necessity of Metaphysics*

In any case, for the sake of our first principle (which should tell us why future men matter by showing that "Man" matters), we cannot avoid taking the imprudent plunge into ontology, even if the ground we can ever hope to reach there should prove no more secure than any at which pure theory must come to a halt. It well may be suspended forever above an abyss of the unknowable. We have intimated before that religious belief has answers here which philosophy must still seek, and must do so with uncertain prospects of success. (E.g., from the "order of creation," faith can argue that according to God's will men should be there in his image and that the whole order should exist inviolate.) Faith in revealed truth thus can very well supply the foundation for ethics, but it is not there on command, and not even the strongest argument of need permits resorting to a faith that is absent or discredited. Metaphysics on the other hand has always been a business of reason, and reason can be set to work upon demand. To be sure, here too, a tenable metaphysics can no more be conjured by a bitter need for it than can religion, but the need for it at least can move us to search after it; and for the search to be unprejudiced, the worldly philosopher struggling for an ethics must first of all hypothetically allow the *possibility* of a rational metaphysics, despite Kant's contrary verdict, if the rational is not preemptively determined by the standards of positive science.

So much in defense of the attempt which is to follow. Only two things we know of it in advance: that it must descend all the way down to the—no longer answerable—final (first) question of metaphysics, so that perhaps from the *meaning* of there being "anything at all" (impenetrable as the sheer fact itself may be) a reason may yet be extracted why a particular something ought to be; and second, that the ethic which may possibly be erected on this ground must not remain bound to the ruthless anthropocentrism which characterizes traditional ethics, in particular the Hellenic-Judaic-Christian ethic of the West. The apocalyptic possibilities inherent in modern technology have taught us that anthropocentric exclusiveness could be a prejudice and that it at least calls for reexamination.

V. Being and Ought-to-Be

Our question is: Ought there to be man? To ask it correctly we must first answer the question of what it means to say of something whatsoever that it ought to be. This naturally leads us back to the question of whether there *ought* to be anything at all—rather than nothing.

1. *Concerning "A Something Ought to Be"*
The difference between the last two questions is not slight. The first, referring to the ought-to-be of this or that, can be answered *relatively* by comparing the alternatives which given reality presents: Since something has to be, then better this than that, therefore it ought to be. The second question, where the alternative is not another being, but nonexistence pure and simple, can only be answered *absolutely,* for example, that being is "good" in itself, for no comparison of degree is possible with nothingness: therefore existence *per se,* no matter what kind, "ought" to be in preference to nothingness, its contradictory (not "contrary") opposite.

The difference for ethics in answering the one or the other question is clearly shown in the example of the introductory question relating to man. One condition of mankind can be found to be better than another and can therewith represent an ought (obligation) for our choice. But over against either (and any) condition one can choose the nonexistence of man, which is certainly free of all the objections to which both alternatives of the former choice are exposed (i.e., as perfect in itself, nonexistence is free of all the imperfections which adhere to every positive object of choice). I repeat: nonexistence can be chosen over all the alternatives of being, but now add: *unless* an absolute preeminence of being over nothingness is recognized.[3] Thus, answering the more general question is of real significance to ethics.

2. *The Priority of Being to Nothingness: How Valid for the Individual*
The recognition of this preeminence and thereby of an obligation in favor of being does, of course, not support the ethical conclusion that the individual should under all circumstances opt for *his* continued existence as opposed to a possible or certain death; that is, that he should cling to his life. The laying down of one's own life for the salvation of others, for one's country, for a cause of humanity, is a decision in favor of being, not for nonbeing. Also the deliberate suicide in order to preserve one's own human dignity from utter degradation (like the Stoic suicide, which is always *also* a "public" act) is in the final analysis committed for the sake of the survival of human dignity in general. For both cases the saying

goes that "life is not the highest of the goods" (Friedrich Schiller). Even the right of individual despair to choose self-extinction, though not uncontested by ethics but granted by compassion, does not negate the primacy of being as such: it is a concession to weakness in the individual case, an exception to the universal rule. The option, however, of *mankind's* disappearing—causing or allowing it to happen—touches upon the question of the obligatory being of "Man," and this necessarily leads us back to the question of whether there should be anything at all rather than nothing.

3. *The Meaning of Leibniz's Question, "Why Is There Something and Not Nothing?"*

This "should be" indeed provides the only defensible meaning to that famed question, otherwise seeming so idle, which Leibniz designated as *the* first question of metaphysics: why there *is* "something at all rather than nothing." For the "why" asked here can obviously not aim at the antecedent cause, which—itself belonging to the things that are—can be looked for only *within* the realm of being and as accounting for parts or states thereof, but never, on pain of contradiction, as relating to the totality of things or the fact of being as such. This logical state of affairs is not altered by the doctrine of creation, which does provide an answer for the world as a whole in the divine causative act—only to revive the question with this very regress, namely, for the existence of God himself. To this, as is well known, rational theology gives the answer of the *causa sui,* self-causation. But the concept is, to say the least, logically questionable; and the glowing confession of faith, "You are from eternity to eternity, God," testifies more to the ultimate logical contingency of a *factum brutum,* which demands ever renewed affirmation, than to an undeniable necessity of thought. We can leave it at that. For even with the positing of a creator, be it necessary or arbitrary, there once more emerges concerning the world (which the puzzle is really about) the question of "why" he created it. Here the religious answer is not the causal one, that the power to do so simply had the deed for its consequence (which would condemn the whole series to brute factuality), but rather that he *willed* it and, if so, then as something "good." (See, e.g., *Genesis* and Plato's *Timaeus.*) But then we must say that this deeming it worth creating was a matter of God's judicious knowledge and not of his blind inclination; that is, that he willed it because its existence is good, not that it is good because he willed it (although the latter was the disturbing view of Duns Scotus). Now, much as the pious is inclined to concur with the presumed divine judgment out of piety and not out of insight into the merits of the evidence, that judgment of worth must on principle *also* be open to independent finding and confirming, that is, on the evidence of the world itself (*fides quaerens intellectum*). In other words, the question of whether

the world ought to be—which it does if its existence (or existence *per se*)
is a value compared with nonexistence—can be separated from any thesis
concerning its authorship, precisely because for a divine creator, too,
such an ought-to-be in conformity with the concept of the good must have
been the reason for creating it: He willed it because he found that it ought
to be. Indeed, it is arguable that the perception of value in the world is
one of the motives for inferring a divine originator (formerly even one of
the "proofs" for God's existence), rather than, conversely, the presupposition of the originator being the reason for according value to his
creation.

Our argument, therefore, is not that only with the decline of faith did
metaphysics have to take over the task which previously theology was
able to perform in its own way, but rather that this task always belonged
to metaphysics and to metaphysics alone—under the conditions of belief
no less than of unbelief: its presence or absence in no way affects the
nature of the task. Metaphysics can learn from theology only a previously
unknown radicalism of questioning, exemplified in the fact that a question
such as Leibniz's would have been impossible in ancient philosophy.

4. The Question of a Possible "Ought-to-Be" Is for Philosophy to Answer

Returning once again to the "why" of the famous basic question of
"why there is anything at all," we found that its understanding in terms
of the causal "from whence" makes the question, in its reference to being
as a whole, nonsensical; but its understanding in terms of a justifying
norm ("Is it worth being?") makes it sensible, at the same time detaching
it from every consideration of authorship, and thereby from faith. Thus
the meaning of the question why there *is* something and not nothing must
be this: Why *ought* there to be something in preference to nothingness,
whatever the *cause* of its coming to be might be? Everything now turns
on the meaning of this "ought."

With or without faith, then, the question of a possible ought-to-be
becomes a task—at least to be attempted—for independent judgment, that
is, for philosophy, where it finds itself at once involved in the question
of *value* and its cognition in general. For value, or the "good," if there
is such a thing, is surely the one thing that of itself urges the *existence*
of its subject from its mere possibility (or, given existence, the rightful
continuation thereof): the only thing, that is, which grounds a valid claim
to being, an "ought-to-be," and makes such being an agent's duty when
it depends on action by him. At this point it is important to see that the
mere fact of value (with its opposite) being *predicable* at all of anything
in the world, whether of many things or few, is enough to decide the
superiority of being, which harbors that possibility within its manifold,
over nothingness, of which nothing whatever, neither worth nor worth-

lessness, can be predicated. Granted that predicability (which we have still to argue), then no temporary or even permanent preponderance of evil over good in the sum of things can cancel the superiority it bestows nor diminish its infinitude. The susceptibility to value discrimination as such, if it obtains at all, and irrespective of its changing and always debatable balance, would constitute the decisive distinction which is not subject to degrees. The capacity for value (worth) is itself a value, the value of all values, and so is even the capacity for antivalue (worthlessness), insofar as the mere openness to the *difference* of worth and worthlessness would alone secure to being its absolute preferability to nothingness. Thus, not only this or that determinate value, when its occasion comes along, has a claim to being, but already the abstract possibility for value in general, as itself a value, has that claim to being and imparts it to the reality harboring such a potential—that is, to the world. This, to be sure, does not tell us why there *is* a world (which remains shrouded in the impenetrable contingency of sheer fact), but might possibly answer the question we substituted for it: why there *ought* to be a world (which would redeem the contingency from issuing in a senseless fact). But all of this holds only if the concept of value is first secured a standing in things and not merely in thought.

5. *Turning to the Question of the Ontological Status of "Value"*

It all, therefore, converges in the focal question whether there *is* such a thing as "value," not as something here and there actual, but as something possible in its very *concept*. Hence, we cannot shirk the task of ascertaining the ontological and epistemological status of value in general, and must delve into the daunting question of its *objectivity*. For with the naked and uncontested fact of subjective valuations at play in the world—that there is lust and anxiety, appetite and revulsion, hope and fear, pleasure and pain, hence also, as their objects, things desired and hated, esteemed and disdained, indeed that there is willing as such and throughout all its choosing of objects the subject's will to its own being—with pointing to this presence of subjective value in the world, little is gained for the cardinal theory and nothing as yet denied to the nihilist. For it can always be doubted whether this whole toilsome and terrible drama is worth the trouble, whether the great enticement is not a great deception. One can always draw up the bill of joys and sufferings: the balance which pessimism—vulgar as well as Schopenhauer's—draws from their presumed totals is familiar and, unproved as it is, can scarcely be refuted on the evidence of the subjective phenomena. On the contrary, contradicting it comes easily under the suspicion of superficiality (or what Schopenhauer called "nefarious optimism"). Indeed even without drawing a balance, we may concede the anguish of willing in itself (not excluding the "will to power," conjured up to replace fallen metaphysics), from which not-willing, and therefore nothingness, would be a redemption. Thus the very intensity of feeling, indeed, precisely the overpowering urge of appetition, can become an argument against their seduction. In short, nothing in the feelings playing their parts in it protects the whole great spectacle from being declared "a tale told by an idiot, full of sound and fury, signifying nothing." And nothing in the fact of its performance prevents the involuntary actors from seeking refuge in nothingness.

It is therefore necessary, where ethics and obligation are concerned, to venture into the theory of values, or rather the theory of value as such. Only from the objectivity of value could an objective "ought-to-be" in itself be derived, and hence for us a binding *obligation* to the guarding of being, that is, a responsibility toward it. Thus, our ethical-metaphysical query about an ought-to-be of man himself in a world that ought to be, turns into the more specific and much less speculative question of the logical and ontological status of values as such. Considering the present, precarious, and confused state of value theory, this is not a promising enterprise. But at least for the sake of clarity it must be attempted. Let us now turn to this task.

3
Concerning Ends and Their Status in Reality

What we must first clarify is the relation of values to ends (or aims), which are often confused with one another but are not the same at all. An end is that for whose sake a matter exists, and which to bring about or to preserve a process occurs or an act is performed. It answers the question, "What for?" Thus a hammer is for hammering; a digestive tract is for digesting and for thereby keeping the organism alive and in good shape; one walks in order to arrive somewhere; a court sits in order to administer justice. Please note that the ends or goals which are here said to define the things or actions do this irrespective of their status as values, and that discerning them as ends is not also approving them. Thus, my stating that y is the end of x involves no value judgment on my part. I may prefer a natural state without hammers to a state of civilization in which nails are driven into walls; I may deplore the fact that lions are not vegetarians and disapprove of digestive systems geared to the carnivorous way of life; I may think it better that people stay where they are instead of always wanting to be somewhere else; I may have a low opinion of every kind of justice dispensed by courts of law—in short, I may declare all these ends to be *worthless* in themselves. Nonetheless, I must still acknowledge them *as* the purposes of those things considered in themselves, provided my *description* of them was correct. Assuming, as it were, the "viewpoint" of the objects themselves, I can then proceed from the recognition of their inherent ends to judgments concerning their greater or lesser fitness to them, that is, their suitability for the attainment of these ends, and can speak of a better or worse hammer, digestive state, locomotive act, or judicial system. *These* then are *value* judgments indeed, but they do certainly not rest upon value decisions or goal-settings of mine: they are deduced from the perceived being of the respective objects and rest upon my understanding of them, not upon my feelings about them. In this manner we can form the concept of a specific "good" and its opposite, and of the degrees in between, for different things and contexts of things: *provided*—and to the extent—that we really can perceive "ends" in the things themselves, as belonging to their nature. It is the "good" according to the measure of fitness for an end (whose own goodness is not judged)— thus *relative* value *for* something.

Here two obvious questions arise at once: *Whose* are the ends which we perceive in the things? And what is the *value* of these ends themselves with respect to which the things under discussion are of value and can be better or worse, as means? Can *they* too be better or worse? The first of these questions aims at the concept of "an end for itself," the second at that of a "value in itself." But because we are initially only concerned with ends and not yet with values, we shall stay here with the first question.

I. The Hammer

1. *Constituted for an End*

Concerning the question "to whom" the end belongs, it must be noted that the expression "to *have* an end" has two meanings. The hammer has the end of enabling someone to hammer with it. It was created with and for this purpose, and this belongs to its—thereto adapted—being in a completely different way than the momentary end of "throwing" belongs to the stone just snatched up, or that of "reaching out" to the bough just broken off for this purpose. The end, we can also say, belongs to the *concept* of the hammer, and this concept, as with all artifacts, *preceded* its existence and was the cause of its origination. That is, the concept here underlies the object, not the object the concept as is the case with class concepts abstracted from, and thus subsequent to, things already existing. There we can ask whether the concept is adequate to its object; in the other case, we ask whether the object is adequate to its concept. The concept of time measurement, for example, was the generative cause of the clock, and the clock is totally defined by this end. It is literally its *raison d'être*. Thus it "has" this purpose truly as a determination of its essence and not merely as an accident of application (which avails itself of a chance fitness). The measurement of time is its destination; indeed, it is so much identical with its purpose that without it it would not exist at all.

2. *The End Not Located in the Thing*

Nevertheless, this concept causing and constituting the being of the clock was not its own but its maker's, who in making it could not also transmit the concept to it. Measuring time was "his" purpose in this authentic sense, remains so, and will never be that of the clock itself. In this second sense of "having," the clock or the hammer does not itself *have* its purpose at all, but only its maker and user "has" it. This is true of all lifeless implements. The purpose *essential* to them *qua* artifacts is yet not *theirs;* notwithstanding their total determination by purpose (or rather precisely because of it) they are barren of purposes of their own.

II. The Court of Law

Let us go to the other end of the spectrum, to the court of law. It too is an artifact, namely, a human institution, and of course here too the concept precedes the thing. It was established *in order to administer* justice. The concepts of justice and its administration underlie the existence of the entity. But not only did the concept here causally precede the thing; it must also have entered into it in order that it might be that *for which* it was created. Brought into existence through "final" causality, a court of law is also held in existence only through this same causality operating within it.

1. *Immanence of the End*
But how can it operate *within* it? It can do so insofar as the acting parts (unlike those of the clock) are themselves animated by the end, that is, if they *will* it and act with a view to it—for which they must be ends-oriented and self-acting beings in the first place. This means that here the difference of being between the maker and the made does not obtain. The maker (e.g., the legislator) and his product (the societal institution) are in kind, if not in person, the same subject, namely, intercommunicating owners of mind and will. Both, therefore, "have" the purpose in the same original sense. At least this is expected, and if, in the performance of the assigned task, other purposes creep in, this too is possible only because the instrumental agency is not simply adapted to a purpose, as an implement is, but entertains purposes on its own; and a deviation from the original purpose (from the founding concept) becomes the occasion for criticism. The criticism would not, as in the case of the implement, fall on the maker but rather on the product apart from him. The clock-maker, not the clock, is to blame if the clock fails, but the judges, not the fathers of the constitution, are at fault if the court fails. The will of the instituting power continues itself in the will of the institution, or else is perverted in it, or modified, enlarged, restricted, and so forth. We need not go into the complicated question of the substitution, compounding, and super-imposing of ends, nor the question of how genuine the professed identification of the officeholders with the intentions of the legislator is, in order to be able to say that ends are at work here all down the line in the same primary sense, if not perhaps always the same ends.

Thus, in contradistinction to the "hammer," it is true for the "court of law" (both are in some sense "tools"!) that a purpose is not only objectively its *raison d'être* but also subjectively the continued condition of its functioning, insofar as the members of the court must themselves have appropriated the purpose for the court to function as a court.

2. *Invisibility of the End in the Physical Apparatus*

Indeed, this subjective side, or the *idea* determined from within, is the *only* thing by which such a societal "tool" can be identified. I can adequately describe the hammer just in terms of its visible form, its composition, the material and shape of its parts, without naming (or even knowing) its end. I can do the same with the clock, even if it is not running. Such a purely physical inspection, through which we can already unequivocally ascertain the respective things in the realm of mere objects, allows us also to recognize their fitness for this or that purpose—and then to assume with extreme probability that they were created with just that purpose: that they were *intended* for it. That is, the invisible ("subjective") intention of the maker is revealed in the visible ("objective") composition of the object; for I know, naturally, that such things do not come about by accident.[1]

But no physical description of men in judicial robes and wigs, in a particular seating order, with a particular sequence of speaking, listening, and writing, etc., can give the least inkling of what a "court of law" is and of what is going on there. I must follow for a while with understanding *what* is spoken there in order to recognize that it is justice and a judicial decision which are at issue, and I must understand these *concepts* themselves to grasp the institution "law court" at all, and then also the event "court session." Only from the totally invisible *idea* of the end (here the idea of justice) do all the outwardly visible signs—robes, wigs, tables, benches, paper, and pens—acquire their meaning as more or less contingent vehicles of its realization. An inventory of the physical items, as complete as it might be, does here not enable us to make the converse inference from means to end.

3. *The Means Does Not Outlive the Immanence of Purpose*

How completely in the case of human institutions the "tool" is not only defined by its "wherefore" but continuously constituted by it can also be perceived in the fact that it is not, like the material tool, an object which, once there, continues to subsist of itself, no matter whether used, and even whether in its purpose understood. The idle hammer may still be found as a physical object after a thousand years and perhaps be recognized. The abolished parliament has vanished into nothing and left behind it only its idea, but no enduring object that one could at some later time put again to the use of its purpose. This is only an extension of the preceding observation. The hammer, even with no purpose named, can be described as a thing that "looks so and so," precisely because its thinghood has an existence separable from its purpose. A legislature, a tax authority, a judicial system have no such independently describable "look," because they have no existence distinct from their purpose.[2]

4. Visible Means Indicating Invisible Ends

Still, even the nonmaterial social instrumentalities make use of material instruments of the first type, from whose discernible utility one can after all learn something of the institutional purpose they subserve. The degree to which this is the case depends on the kind of purpose involved: the more this includes physical activity, the more it will be inferable from its physical means. From the potsherds of the Athenian vote for ostracism, even from seeing them distributed, collected, sorted, etc., an extraterrestrial visitor with nothing but these visible data could learn utterly nothing (as little as from a modern ballot) about the meaning and intent of the institution. But from the arsenal of the Athenian army of the same period he could learn a good deal, not to speak of its use in battle. A modern arsenal, its technical ends decipherable from its physics, would yet more unmistakably speak for itself. Nevertheless, even in so clear a case, the entity "army" is not really knowable, in its societal and political essence, by its physical means ("hardware") nor by its physical actions, that is, by the outer aspect alone, and here, too, the inner conceptual purpose of the whole and its parts (e.g., a knowledge of what "the state," "sovereignty," and "national conflict" are) remains the ultimate datum by which to understand the social means. In most cases this purpose is highly abstract and beyond disclosure through the physics of its means: not the most elaborate torture tools of the Inquisition can tell us what a trial for heresy was about, nor all the exchanges of papers and signatures, what a contract is, nor what "property" (if this happens to be its object). And not even the clearest purpose of atomic weapons in their eventual use— namely, annihilation—betrays that the purpose of their being stockpiled is that they rather not be used.

5. Man the Seat of Purpose

Let us here break off the rather elementary discussion of these two classes of purposive entities and sum up what we have learned from them. They were, as should be recalled, the opposite ends of a spectrum exemplified by "hammer, digestive organs, walking, court of law," and were picked out as the first to compare because of their contrast in one respect and their likeness in another. They are alike in that with neither can there be a doubt that it was *made* for a purpose and with intent suited to it, so that here most clearly the fitness for it is no accident. Both, in short, are human artifacts, and from this it also follows that the ends respectively entrusted to them are human ends, namely, particular ends of their makers and users, be they individuals or society. At the same time they are utterly different in that the end is external to the material implement, while to the social instrumentality, composed *of* as well as *by* men, it is internal in the sense described. Even so, the latter too remains a means and does not, with this interiority of the end, itself become an end. We have thus not yet arrived at an "end in itself," that is, something being *its own end*. This, in the two cases, would be the human subjects whose ends they subserve; and if this were the whole picture, then the end-in-itself would always be man or would reside in man, which would make man the ultimate end in general. This would be in keeping with the modern conviction that "purpose" as such is an exclusively human phenomenon, existing nowhere outside of the mind of man and only by him conceptually imparted to other things—to his artifacts in the instrumental sense terminating in himself, to other creatures (animals) in the terminal sense, "anthropomorphically" transferred from himself.

Be this as it may, of the entities just considered it is true, both that they are unequivocally purposive entities, and that the purpose is set and entertained by *human subjects*. We only add that for an entity to be "unequivocally a purposive one" does not necessarily imply that the purpose itself is unequivocal. Several ends can be fused already in its original conception or be added in its later functioning, or creep in, etc., even to the point where the original end is completely replaced by others; and for the societal-personal instrumentality, the office, it is not at all impossible that, contrary to every original intention, it makes itself an end in itself.

III. Walking

1. *Artificial and Natural Means*
     The two remaining examples concern nonartificial, that is, natural things and functions. We must examine what bearing the distinction between artificial and natural has on the aptness of things to have "purpose" attributed to them. In addition, within this class of things natural, represented by "digestion" and "walking," there enters the distinction between voluntary and involuntary function, which divides the two cases, even though the "tools" in question are equally gifts of nature: this distinction too affects the question of attributable purpose, and we see at once that it cuts across that between men and animals. Thus we are drawn beyond the domain of purely human intention, perhaps even that of intention and design (strictly speaking), in several respects: with the natural givenness of the organ in all cases, and with its function—whether involuntary or voluntary—in the nonhuman cases. We shall begin with walking, or locomotion in general, as an example of the "voluntary" class, in which human intention at least has a place, since the moving organism may be a human one.

2. The Distinction between Means and Function (Use)

We said, "One walks in order to arrive somewhere." The "in order to to" designates the purpose. One walks "with" the legs; these (together with the whole, associated neuromuscular apparatus) are the means. The means is given by nature and is alive, a part of the living user himself, but does not set itself in motion, and possessing it is distinct from employing it. Not the legs walk, but the walker walks with them; not the eyes see, but their possessor sees with them; and the "in order to" also indicates, besides the purpose, a control on the part of the subject, which we call the will. This at any rate is the assumption with regard to external motor functions generally, or at least (on the strength of subjective evidence) in the case of men. Sensory functions are less voluntary. One sees, hears, smells without willing it. Sensation has always been understood as a receiving or suffering ("being affected," "receptivity"). But even here, the active and voluntary element can come into play: Looking is more than passive seeing, listening more than hearing, sniffing more than just smelling, and so on. With tactile "feeling," which indeed *per se* includes motor activity, the active-volitional share in perception is evident. (Actually, even though largely involuntarily and unnoticed, motor function is involved in the most unintentional act of vision, e.g., in the focusing of the lenses.) Here, then, the "in order to" as subjective purpose enters into the use of the sense organ as well.

However, the clearest case of the distinction between possession and use of the organ, and thus between its own, general purpose and the particular ends of its employment, is our motor equipment, where we say of him who is endowed with legs ("walking-tools") that he is at liberty to walk or not to walk, and if he walks, to walk here or there. All this may be dictated to him in other ways, but it is not dictated to him by the possession and the capability of the legs. "He is at liberty," therefore, says nothing yet concerning his freedom in general, but it says that *by the organ* it is left to "him" (whatever this subject itself might be) whether and what for to use it—that he and not the organ decides about this.

Here we are thus closest to the hammer. It too is only there *ready* for use, but not causing it. There is no need to puzzle over freedom or determinism of the will itself, to be right in saying that the tool can be used "at will"—where the willing may very well have its own determination. And this also the natural tool of the "voluntary" type has in common with the artifical, that the purpose of the tool does not yet specify the purposes of its function. That of the tool is simply the function itself, but this, being voluntary, must have its own purpose set to it in order to be undertaken, and only rarely is the function as such that purpose. Legs fulfill their purpose in walking as hammers do in hammering, but whether these fulfill *their* purpose is another question. Indeed, the tool has done its duty even when used aimlessly or foolishly and to no avail. One can

also say that the purpose of the tool (or organ) is general, that of its
employment, specific: the specification of the general capacity is imparted
to the action by its particular purpose (and also, of course, by the external,
physical circumstances).

3. *Tool, Organ and Organism*

This similarity of limbs (external motor organs) with tools was indeed
the reason why the former—and then by extension *all* of the body's
functional structures, internal no less than external, sensory and chemical
no less than motor—were called "organs," which means precisely "tools":
something which performs a "work" (= root *erg* in *organon*) or with
which a work is performed. Aristotle in his famous definition of things
alive *defined* the living body straightway as "organic" (*soma organikon*),
that is, a body endowed with, or composed of, tools. And for good reason
he called the human hand the "tool of tools," both because it is, so to
speak, the exemplary tool itself, and because with it the artificial tools
are made and employed ("handled") as its extensions.[3] So if, in speaking
of "organism," we were keeping to the original, literal sense of the word,
we would already be speaking of a purposive entity, for "tool" cannot be
thought of without the idea of "purpose." But the success story of a name
naturally proves nothing concerning a substantive issue, and whether
natural tools, like artificial ones, have purpose already at their origin and
in their being, apart from being employed thereto, is as yet undetermined.
We shall have to deal with this question, whose importance will be seen
later, when coming to the last class of examples (represented by "digestive
system") where the working of the tool is as little willed as its existence,
indeed largely coincides with it, and where therefore the question of
purposiveness cannot have recourse to any subjective setting of purpose.

4. *The Subjective End-Means Chain in Human Action*

Even in the present class of examples (represented by "walking"), the
role of subjective purpose is by no means undisputed, despite the action
being voluntary. That one walks "in order to" get somewhere is surely
convincing where the subject is human. For there one can ask the walker
why he walks and receive a whole series of answers, of which the naming
of the "whither" is only the first, and each further "in order to" leads
into another: in order to get there—in order to meet a friend—in order to
discuss something with him—in order to come to a resolve—in order to
thereby discharge a duty . . . and so on. Of each object in the sequence
it can be said that it is *willed* at its place, and, in fairly rational action,
that the later is willed before the earlier, that is, this for the sake of the
later. We can leave undecided whether the series really (as it has most
often been portrayed since the ancient philosophers) must always terminate in a definite final object as the true "for the sake of which," or

whether its ever-lengthened projection doesn't rather lose itself in the general labyrinth of existence; also, whether the conscious reasons are always the sole or true ones; whether the distant object of the will in fact always determines the closer as its means, or perhaps, conversely, the proximate object generates the fiction of a more distant goal. No matter whether the whole linear image of a series is not merely an idealized model to which in reality corresponds an intricate web of many strands: notwithstanding all these unclarities, it is still clear that the action-texture here considered is really purposive in the subjective sense, that is, governed by preconceived goals (from here, after all, the whole concept of goal and purpose stems), and that here the "for the sake of which," the more or less distinct intention, indeed tells us what the train of events is about.

5. *The Objective Means-Ends Mechanics of Animal Action*

It is just as clear (or almost so) that no such articulated end-means chain can be imputed to the action of animals, though surely it too is *goal*-directed. Surely, the bird couple gathers stalks *in order to* build a nest, and later worms, in order to feed the nestlings, but no one would assert that the absorption in the first goal looks already beyond it and has, across the weeks to come, the second "in its view"—together with all that is to happen in between: egg-laying, brooding, hatching, etc. Rather would one say here that the entire series grows step by single step "instinctively," out of an obscure urge, which at definite times, occasions, or cues compels the appropriate (preprogrammed) steps, each set on satisfying *just itself* and in this sense "blind" (namely, for anything beyond itself), yet eminently "seeing" within its own scope, insofar as its execution involves utmost discrimination of sense and control of motion—and certainly imbued with feeling and "willing," as can be seen in the passionate agitation shown especially in the face of obstacles. This mixture of "seeing" and "blind" is a puzzling matter which is little aided by the magic word "instinct." No observer can overlook the enormous involvement of "interest" which permeates much of the "voluntary" behavior in the more alert species of animals with central nervous systems: the profound emotionality of their goal pursuit in matters of food, sex, and rearing of the brood; their utmost yes and no in physical threat and self-defense. And yet at the same time the observer must disown this whole impression of his insofar as his theory denies to these subjective occurrences a guiding and goal-setting force and views them (i.e., their physical side) as merely links in the necessitarian chain of stimulus and response, which in *all* of its steps is determined by objective causality alone. Here various levels of interpretation are possible.

a) When the cat watches the mouse we can say that it does so "in order to" pounce upon it at the proper moment; and when it pounces, that it does *that* in order to kill it, and kills in order to eat it, and eats in order to quench its hunger. But this is not saying that the cat set out to watch the mouse in order to quench its hunger—if, that is to say, the "in order to" is understood as *anticipatory beholding* of the ultimate purpose. We rather suppose the anticipation (if present at all) to reach just to the immediate, proximate goal (the watching just to the imminent pounce, etc.), but not to the mediate goals further on and up to the final goal: this, the overall goal of the whole series, results only additively from the succession of single steps. The first restriction upon animal behavior even in the cerebrally higher species, compared to human behavior, would thus be the confinement of subjective anticipation—of what is "known" and "willed" in it—to the momentarily nearest goal, so that the whole purposive series is pieced together of single subpurposes, each one leading to the next.

b) But the overall purpose, here the quenching of hunger—by what, in the absence of a knowing will, was it instilled in the behavior train? The natural answer here is: precisely by the hunger, which—taken for now as a feeling—sets the whole train in motion and dominates it. Since this is a feeling which, as "suffering," has an inner urge toward its own elimination, one can say that *this* is the unifying subjective motivation of the entire sequence, even without anticipatory goal image and a choosing of means conducive thereto. *With* this urge behind it, the series is initiated and carried through "in earnest." Without it, the single acts can also be executed independently, by way of reflex or "in play," and are then easily diverted by other stimuli, as can be frequently observed. Thus, generally speaking, feeling associated to need is the psychic proxy of purpose in the voluntary behavior of prerational life.

c) Feeling, however, only drives onward and teaches nothing concerning the means conducive to its pacification. Whence, then, comes the appropriate animal behavior in its single steps and their often extensive sequence (nest-building, and so on; selection and stalking of prey, and so on)? Here the answer is that the smaller action-complexes, of which the longer chains are composed, have their ready-made schemata in the programming of the organism, through which the impulse of feeling is channeled. The schema comes into force at its own cue (external or internal stimulus), and its execution has the same urge and satisfaction of feeling in its small scope as the whole chain has in its larger scope. The purpose thus dwells on the one hand in the impulse, on the other hand in the preimprinted behavioral form ready for it.

d) But if in this way the sequence is in truth more driven forward by the imperative impulse of feeling than guided by the projected goal of its satisfaction—rather *from* hunger than *for* satiation—and its single steps are more triggered by the "cue" than inspired by the end, then the position of purpose in this progression of events becomes rather cloudy. For the feeling of hunger, so we are told, is nothing but the psychic equivalent (or even less: the symbolic, surface appearance) of a physical deficiency in the metabolic system which also, through its own chemical and neural mechanisms *beneath* feeling, produces those bodily states (purely physical again) that *really* cause the motor behavior, complete with its emotional garb. If this, however, and not feeling (demoted to a mere symptom) is the real cause of behavior, then "purpose," if it is at all to play an effective and not merely decorative role, must have its seat *first* in *this* causality and not in the feeling mirroring it—and in that case has been allowed an existence outside the psychic realm in general.

This raises the question whether purpose exists in the objective, physical world or only in the subjective, psychic world—a question which we reserve for the last class of examples. But whatever the answer to this ontological master question, the question concerning the status of subjectivity confronts us right here, and most clearly in connection with those *single* action steps we found in their own way to be "seeing," whereas the emotion propelling the whole action sequence is certainly blind regarding both final purpose and intermediate links. The aforementioned piecemeal composition of the long purposive series by short, self-enclosed part purposes, each having its direct object present in perception and will, bids us to explore these single links, that is, the sharply intentional units of behavior.

e) We say that the cat eats the mouse to quiet its hunger, kills it so it can eat it, and stalks it so it can catch it. But perhaps we should rather say: It devours the mouse *out of* voracity (not in order to be sated), kills it out of the lust for killing, catches it out of the lust for catching, and stalks it out of the lust for stalking. The result is the same, but the explanation is different. For in the latter case the felt greed or lust can be viewed each time as a mere secondary, subjective appearance (psychic token) of a physiological state of stress, such as we conceded in the case of the all-pervading feeling of "hunger," and this *physical* precondition in the organic base together with the sensory "cue" would arguably be enough to *trigger* the appropriate behavior pattern. Thus, for example, the mouse must certainly first be sighted (cue!) for the attitude of stalking to be taken up, and then the stalking must constantly take account of the visible movements of the prey if it is to achieve its object: to this extent it is correct to say that the process is a seeing and not a blind one—that it has the goal "in sight." But how seriously is "seeing" to be taken here in the sense of mental awareness? In terms of a cybernetic explanation, the sighting can be understood as a mere, objective nerve stimulus acting as a "trigger," and the behavioral adaptations which follow, as an equally objective sensorimotor feedback mechanism—so that the whole sequence of events is describable roughly as follows: physiological stress (homeostatic "gradient"); internal secretion and nerve stimulation; readiness for

selective triggering of a behavioral pattern; external (sensory) nerve stimulus triggering the appropriate behavior pattern; sensorimotor feedback
guiding the course of behavior; success of action eliminating the initial
stress (homeostatic equilibrium), this itself being the "goal." Hence the
following back-and-forth in our example: mouse sighted—stalking, mouse
in proper position—pouncing, mouse in claws—tearing, mouse torn apart—
devouring; the general condition underlying all this: a metabolic lack
(dishomeostasis) with its internal reporting and excitation system =
"hunger."

According to this explanation, every exertion of animal life would have
but one goal and that a negative one, namely, alleviation of stress; or
rather, since the word "goal" has become inappropriate here, the course
of all animal action follows the law of terminal equilibrium—the mechanics
of entropy. The *bonum desideratum* would be the subjective representation of the state of indifference or quiet which waits at the end. What
passes itself off for effective striving would be the unidirectional drift
toward easing of tension, the pleasure felt at attaining it would be the
fancy trimmings of a disappearing act, that is, of tension ceasing in a
(momentary) condition of rest. What actually happens in this train of
events would be only indirectly represented, not truly described, by these
accessories of feelings and mental images: the mental is secondary, the
physical primary.

f) One will note that in the purported "true" description no psychological expressions (subjective concepts) appeared: homeostatic gradient,
not hunger; sensory stimulus, not seeing; ceasing of tension, not satisfaction. We ask: What role in the *purposive* character of the action-context
can we then still attribute to the subjective, whose concomitant *presence*
as such no witness of the expressive intensity of animal acting and suffering can possibly deny? As is well known, the dogmatic violence necessary to deny it nevertheless has not been lacking in the history of theory.
But the totally artificial rationale for such a denial, namely, Descartes'
decree that subjectivity as such can be only rational and therefore only
exist in man, does not bind the reasonable observer, and every owner of
a dog can laugh it off.

If, then, the presence of the subjective as such is indisputable (no matter
where in the evolutionary chain it begins to appear), the question for us
is what it signifies. Does it, for example (as psychophysical parallelism
and materialism assert in different ways), signify no more than an accom-
panying tune without influence on what it accompanies? It is the question
concerning the power or impotence of the mental, here narrowed down
to the question of the influence or lack of influence of subjective *purpose*.
Its presence is conceded with the admission of subjectivity in general,
with its pleasure and pain, pursuing and avoiding. But is the purpose thus
"experienced" also an effective factor in the bodily behavior? It must be

noted that this question only concerns the *relation* to the physical, not the nature of the physical itself. In other words, the question of whether or not purpose can be found *there* (in a nonsubjective sense) is untouched by the resolution of the question of whether *mental* purpose has or has not *trans*mental force. The latter question came up, it will be remembered, in connection only with the *optionally* movable limbs of the body: we are still dealing with the example of "walking" and not yet with "digesting." The adjective "optional" must here not be so understood that it prejudges the answer, namely, as saying that the will governs the limb. ("Voluntary" is avoided because it does seem to beg the question.) Not even there being a mental will at all (still less its being "free") is necessary for the distinction here made between automatic and optional operation of organs. The distinction, as said before, signifies at first no more than that the operation of certain organs—the external motor organs—is variable, occasional, and under central-neural control, while the operation of others is given with their possession and is automatic: for example, clenching of the fist on the one hand, beating of the heart on the other. But at present, with the assumption of subjectivity in general, we have already assumed the will (or its animal analogue), and our question now is its role in "voluntary" movements.

And now we may finally add (to the deserved discomfort of those who find it easy to make their peace with the asserted nullity of "soul" in all of nonhuman nature) that the answer to the question in the animal case covers the case of man as well—that is, that the distinction made earlier between the multipartite end-means chain of thinking man and the single-step goal-apprehension of the feeling animal becomes irrelevant in the face of the basic question concerning "power or impotence of subjectivity" in general. For what we said in *e* about the immanent cybernetic sufficiency of the purely physical series in animal action can, with adequate refinement, be transferred to the motivational, thinking, and decision-making texture of man's inner life as well, even unto the highest reaches of reflective consciousness, which after all has still its cerebral (= physical) basis. There too, according to the theory, the physiological (neural) events, once completely known, would suffice for a causal explanation of the outwardly visible, that is, *physical* "behavior" (in the most comprehensive sense that includes verbal utterance). The inward aspect experienced in consciousness, which passes itself off as the explanation to the subject, would be, causally considered, nothing more than an idle ornament—and a deceiving one at that. Indeed, only the most exorbitant of metaphysical *ad hoc* assumptions (from which, to be sure, speculative thought, capable of everything, has not shrunk) can make man a sole exception to the rule, if in all the rest of the living world the "subject" is nothing but an ineffective accompaniment and thus in its own testimony mere appearance. The status of subjectivity there must affect the status of human ends as well, and hence that of ethics. We shall see later that, in addition, it also affects "downwards" the question of purpose within nonconscious life and, still further down, in inanimate nature—that is, in the world as a whole.

At this critical point, if we really wished to do justice to the systematics of our theoretical enterprise, we should interrupt its progress with a full-dress discussion of the causal role (or nonrole) of consciousness in general—that is, reenact the whole stubborn "psychophysical problem" which has plagued philosophy since the time of Descartes. In order not to burden the present discourse with such an extensive digression, over which the reader might lose the thread of the main argument, we have relegated this subsidiary discussion to an Appendix on "Impotence or Power of Subjectivity" (see pp. 205–231). The reader is asked to assume its result here as a supposition in what follows.

6. *The Causal Power of Subjective Ends*

The result assumed from here on and briefly put is the restoration to credibility of the original testimony of subjectivity in its own behalf, that is, a vindication of its prima facie claim to "reality" against its denial and the demotion to a mere "epiphenomenon" by materialism. The reality thus granted is as "objective" a fact in the world as is that of things corporeal. In its case, too, reality means efficacy, namely, to have causative force inside *and* outside of itself—thus: for thought to have the power to determine itself in thinking and, through this, the body in acting. But with the determination of the body, which hence continues forth into the surrounding world, subjective purposes acquire an objective role in the fabric of events: that fabric, therefore, that is, physical nature, must have room for such interventions by a nonphysical agency. The long-held, would-be axiom that nature does on principle not allow this room is an overstatement of its determinism, which the most recent physics no longer shares. So much here for summing up the findings of the lengthy Appendix. Concerning the method of its argument, we only add that it is in the main negative: the contrary proposition, that subjectivity is powerless, is reduced *ad absurdum* in its logical, ontological, and epistemological consequences; moreover, it is shown to be unnecessary for its intended purpose, namely, to preserve the integrity of the laws of nature. The showing of this nonnecessity leads beyond the merely negative argument in that it tries to exemplify the *compatibility* of psychophysical interaction with the validity of the laws of nature in a positive, if entirely hypothetical thought model. All that is claimed for the model is possibility, not truth, that is, that it neither contradicts the phenomena nor itself. But since alleged impossibility—to reconcile the controversial interaction with the causality principle of physics—was the sole reason for the *tour de force* of parallelism or epiphenomenalism, the mere instantiation of possibility in a thought experiment suffices to render unnecessary the escape of despair into viewing mental life as powerless and a sham, and thereby to rob it of its only excuse. In this way, the basic experience of feeling life is restored to its birthright, simply because no theoretical duress forces thought any longer to the monstrous alternative. (The duress, by the way, was never more compelling than its—essentially unprovable—premise: absolute determinism itself.) The "soul" and hence the "will" is vindicated as a principle among the principles of nature, and this without our having recourse to dualism (a recourse not quite as desperate as that to materialistic monism but still highly unsatisfying to theory). Thus, we can say with some confidence that the realm of voluntary bodily movement in man and animal (exemplified by "walking") is a locus of real determination by purposes and goals, which are objectively executed by the same subjects that subjectively entertain them. Implicit in this statement is the recognition that efficacy of ends is not tied to rationality, reflection, and free choice—that is, to men.

The efficacy *is* tied, however, as far as our argument has gone, to "consciousness" in some sense, to subjectivity and to "volition," for which our examples were chosen. So the question arises whether also beneath this, on the unconscious and nonvolitional levels of life (not to speak of inanimate nature underlying this in turn), something like a "purpose" is at work. We now turn to this question, which is fundamental to an ultimately ontological grounding of "value" and thereby of ethical obligation, although we dare not expect nearly the same certainty of answer to it as the preceding question allowed (and besides will have every bias of modernity against us).

IV. The Digestive Organ

1. *The Proposition That Purpose in the Physical Organism Is Only Apparent*

a) Every organ in an organism serves a purpose and fulfills it by its functioning. The overarching purpose jointly served by all the special functions is the life of the organism as a whole. That this has itself, that is, its own life, for a purpose can be asserted in several senses. The most neutral and least controversial of these is that in an organism everything is *de facto* so arranged that in effect it contributes to the maintenance and performance of the whole, just as in a machine everything is arranged so that it contributes to its total functioning. Stating matters thus asserts nothing yet about the kind of causality at work, for example, whether it is teleology ("final causes"). For the machine we know that this is indeed the case in its manufacture but not in its functioning. Regarding the organism, the reigning theory holds that it is not even the case in what here corresponds to "manufacture," in its genesis. "Genesis" has here a dual sense: growth of the individual (ontogeny) and origination of the species (phylogeny). Ontogeny is understood as the causally necessary execution of the genetic determinations ("program") contained in the seed, for which no teleology (entelechy) need be invoked. Phylogeny is explained by the concurrent mechanics of the random variations in such determinations and the natural selection of their results—thus likewise excluding all teleology. For the "working" of the resulting structures, this exclusion holds anyway: coming-to-be, arrangement, and function—all of them—only look *as if* they were governed by ends. According to this view, the organism would be even less teleological than the machine, whose manufacture at least was teleologically determined.

b) Now, this teleology of the machine, borrowed from without, has its original seat in the producing organisms, the human constructors, who accordingly cannot themselves be of a wholly nonteleological nature. But *their* teleology, so we saw, is held by reigning theory to exist in their ideas only, and not in their causative action itself: this, too, only *looks* as if it were governed by subjective ends (indeed, even the ideas only look that way, according to epiphenomenalism!). Subjectivity of purpose is thought to somehow go together with the objectivity of purely causal action without imparting to it more than the appearance of teleology.[4] The subject, to whom his own thinking and acting "look" that way, is the same to whom—for this very reason—also the being and acting of organisms in the world look that way.

2. *Is End-Causality Confined to Beings Endowed with Subjectivity?*

This "as-if" and sham conception of subjectivity I believe to have refuted in the Appendix. But with the vindication of subjectively determined action, *purpose* in living things is vindicated only as far as "consciousness" extends—that is, only for species endowed with this, and in them again only for such of their actions that depend on it and especially on will ("voluntary" actions). But what about digestion and all the other subconscious, nonspontaneous organ functions in these same species? And what about the total life of unconscious (noncerebral) organisms in general? If we stopped at the point we have reached, we would be left with a strange dichotomy, which is in itself not impossible. We would have to say that with the evolutionary appearance of subjectivity an entirely new, heterogeneous principle has entered nature or come forth in it; and there would be a radical (not just a gradual) difference between the creatures that partake of "consciousness" (then, in degrees) and those who do not; and even within the partakers themselves, between those of their activities which are subject to consciousness (or partially so) and the much broader range of activities that are not.

a) *The dualist interpretation*   As we said, such a dichotomy is not unthinkable, and it can be thought of in two ways, as in fact has been done.

Either one thinks, dualistically, the alien principle (the subjective "soul") to seize upon and make use of configurations of matter (brains) which favor it but came about for other evolutionary reasons and not "for" it (unless soul had a hand already in bringing it about); in that case, the adventitious (transcendent) principle would have ingressed *into* nature *at* the opportunity offered. *Or* it is thought, monistically, to have, *with* the opportunity, come forth by immanent logic *from* a nature which had reached this point. The first, dualistic alternative is clean in itself, but suffers from all the objections which dualism as such has against it. One such objection follows from the gradualness of transitions and the minimal magnitude of evolutionary beginnings: it is awkward, not to say grotesque, to carry dualism, and with it a share in transcendence, into the amoeba or wherever else "feeling" begins.[5] The ultimate stumbling block, however, is this: the theory of "ingression" implies the soul (or whatever it is) to exist beforehand and, as it were, to wait in the wings for its opportunity to enter the stage. In other words, it is held to enjoy an independent transcendence which only for its physical manifestation depends on specific physical conditions. But let us call to mind what is probably the strongest empirical argument on the side of materialism: that by every evidence of human experience, matter occurs without mind, but not mind without matter, and that no example of a disembodied spirit is known. Just this, however, that is, an independent realm of effective, immaterial transcendence, must the theory of ingression assume; and that, although logically unimpeachable, is the inductively least supported and ontologically most outré of all conceivable hypotheses—which has not prevented it from being the most powerful one in the history of thought's struggle with the riddle of the soul.

b) The monistic emergence theory   So there remains the other alternative—that soul and mind emerge *out of* nature when, not quite accidentally, the suitable material conditions occur: subjectivity emerges *with* those novel conditions as a concurrent, complementary modality of their own being. This in fact is the bearing on our problem of Lloyd Morgan's (and others') theory of "emergent evolution," according to which new, more comprehensive causal structures—like atom, molecule, crystal, organism . . .—superimpose themselves on previous levels which did not indeed foreshadow them, but by leaps pass over to them whenever certain critical threshold states of organization are reached.[6] The real qualitative *novelty* of such "emergents" is emphasized at once with their nontranscendence, that is, their strictly immanent origin. In the case of consciousness this is a valiant attempt to have the advantages of dualism (recognition of the irreducibly own nature of the new level) without the disadvantages of its metaphysics (having to posit a substantive "beyond" of matter). In terms of our problem, the proposition allows us to regard the appearance of subjectivity as one of those evolutionary "leaps" and to keep the conception of the preceding levels that underlie it uncontaminated by the imputation of "purpose," this being peculiar to the new level alone. Then, indeed, conscious action would be purposive, just as we assert, while unconscious organ function (with which our question now deals) would be not. This clean division is just what "qualitative leap" implies.

The theory is tempting; but if the dualistic one was ontologically awkward, this one is logically so. For one thing, here too the gradualism of transitions conflicts with the image of leaps. Moreover, the difference between the realm of material things *in toto* and that of subjectivity (or, the leap from one to the other) is of an altogether different order than that between, say, atom and molecule, or between free amino acids and the first cell, or between unicellular and multicellular organism: all these are still of the same, physicochemical order, and no functioning of a lower level is altered by the supervening higher one. The truly *generic* distinction of the "subjective" from *all* this is not really comparable to the gradations of complexity within it. Most of all—and here comes the logical flaw— the emergence theory, when embracing this generic leap, buys self-consistency with ignoring the *causal* angle. If the new principle is to have *power* (an essential attribute of it, as we saw), then its supporting base— the simpler level—must comply with the evident condition that nothing can give birth to what is *entirely* alien to it and runs counter to its immanent law, and thereby does violence to itself. (Or, one would have to adopt a "dialectical"—Hegelian—concept of being, in particular of "matter," which the teachers of emergence most certainly did not have in mind.) A theory operating with mere degrees of (material) *complexity* and their ensuing morphology cannot encompass mind, which falls outside these terms. Were consciousness nothing but a new *quality* added to, and somehow superimposed on, the preceding one, that might pass as at least innocuous, though not quite intelligible. But we found that it is also a new *causality* which reacts upon its base and changes its course of action. The physical things which come under the influence of subjectivity no longer run as they would without it. The new level thus has the power to constrain the substratum, out of which it emerged, at any rate to codetermine it. This is inconsistent with the idea of emergence, in which the new *adds* itself to the old without changing it—rather *supplementing* it as an expression of the level of organization attained in it. A mere quality could do this, but it must be causally innocent; that is, it must, despite the novel *forms* of action in which the more complex causal system of the substructure now expresses itself, not become itself a *factor in* this system. The theory only explains new action patterns, not new actions. For instance, the purposiveness of consciousness, foreign to the substructure, would have to be confined to consciousness itself and not reach back, as a transitive force, into the substructure. *Domination* by consciousness is herewith precluded. But this is tantamount to saying that the emergence theory, if it seriously holds to the essential novelty of the top layer, must ally itself with some form of psychophysical parallelism or epiphenomenalism—generally with the proposition of the impotence of consciousness as a pure quality; and this we have refuted.

Or else it must say: What looks like a leap is in reality a continuation; the fruit is presaged in the root; the "purpose" which becomes *visible* in feeling, willing, and thinking was already present, invisibly, in the growth leading up to its manifestation: and that not just in the sense of a permissive openness to it in case it should one day ingress into physical causality from above, but in the sense of a positive predisposition and selective tendency toward its eventual manifestation, should conditions open the way for it. The growth, then, was really *toward* it. In other words, the evolving antecedent must be credited with a positive potentiality for the novel mode (not *totally* novel, indeed, by this view) whose time would come. That coming must be understood as actualization, as "telos," as fulfillment of a movement tending toward it. In short, only in connection with a generally "Aristotelian" ontology is the theory of emergence logically tenable. But this is precisely what the theory wished to avoid: the substructure was to be spared the need of being interpreted with a view to the superstructure; there was to be *no* need to import categories of explanation from the latter into the former; the newly apparent causality of the one was *not* to be seen as foreshadowed in the other and aiming at this appearance. In short, teleology was to be avoided. However, that leads, as we showed, to the dead end of the absolute leap and of the impotence of mind.

Thus we can say that the—theoretically valuable—principle of emergent *novelty,* if it is not to be totally arbitrary and hence irrational, must be tempered by that of continuity: and a substantive continuity, not a merely formal one—so that we must let ourselves be *instructed by what is highest and richest concerning everything beneath it.* This, however, is no marginal correction but rather strikes at the heart of the matter. As one knows, today "continuity" means conversely that one learns from the lowest about everything above it! It was just this reductionism that the emergence theory sought to escape without having to opt for the opposite course, thus avoiding a troublesome choice. But here one must say, *non datur.*

3. *Final Cause in Preconscious Nature*

With that we have indicated our own position. Reality, or nature, is one and testifies to itself in what it *allows* to come forth from it. What reality is must therefore be gathered from its testimony, and naturally from that which tells the most—from the most manifest, not the most hidden; the

most developed, not the least developed; the fullest and not the poorest—
hence from the "highest" that is accessible to us.

a) *The abstinence of natural science*   This testimony of our own being
is consciously ignored by the natural sciences because of the well-grounded
prohibition of anthropomorphism, also because of Occam's principle of
parsimony, and finally just because "goals" are not quantifiable—and that
is methodologically in order. The biologist investigating elementary life
processes, for example on the molecular plane, proceeds *as if* he didn't
know that there was a whole organism in which they were occurring;
investigating lower organisms, as if he did not know that there were higher
ones; investigating the higher, as if he did not know that they possessed
subjectivity; investigating the highest (and his brain), as if he did not know
that thinking determined his being. That is, he places himself at the point
of those "beginnings," at which indeed no one but God could foresee
what would one day evolve out of them, or at the point of those elemental
components in the evolved entity itself, from which again no one but God
could discover what invisible complement (e.g., awareness) goes with
them. And so it behooves human science.

b) *The fictive character of the abstinence and its self-correction by the
scientist's existence*   But it behooves this science no less to bear in mind
that this is a fiction. The methodological utility of the fiction is evident
and need not be discussed here, because it is not as such called in question.
But methodological advantage must not be confused with ontological judgment. Naturally, the researcher who is busied with the beginnings of life
does know about the whole evolutionary chain, the one busied with cell
metabolism knows about the whole organism, and the one who is busied
with the brain knows about thinking. Indeed, from this knowledge alone
stems his *interest* in the investigation of the elemental. Above all does he
know about this interest of his and about his thinking devoted to it. This
he must take seriously in its automony, else he could not hope for the
attainment of truth, or even for so much as distinguishing it from falsehood, and ascribe to his thinking any sort of validity. But in presupposing
the autonomy of his thinking, that is, its inward power, he has logically
(as shown elsewhere) already acknowledged its outward power as well,
and hence the power of the motivating interest, too, since mental self-
determination is only possible in union with spontaneous body-determination—thinking, only with internal *and* external freedom: for example,
in writing down the results of his thinking. But with this (if he doesn't
take refuge in a fantastic dualism) he has recognized mind, indeed subjectivity and *interest* in general, to be an efficacious principle *within* nature—thus implicitly broadening his concept of nature beyond his asserted
model. Taking himself seriously (as he must) and at the same time not as

a unique exception (as he cannot do as a member of the human community) he cannot but give nature credit for bringing forth goal-causality, and hence regard the latter as not completely foreign to the former. He may then, in the analysis of pure matter (an abstraction from the *plenitudo entis*) still stick to the chosen, purely "external" minimum account, as the business of physics requires; he only must resist the temptation to turn the artificially reduced minimal data into a reductionistic metaphysics—a temptation apparently more difficult to resist than the opposite one of anthropomorphism.

c) The concept of "ends" beyond subjectivity: how compatible with natural science   To the philosopher it still remains to show what it means for the status of *purpose* that the subjective evidence for it bears beyond the domain of subjectivity on the concept of nature as a whole. It should be noted that we are interested in the concept of nature for the sake of a theory of purpose, not in the concept of purpose for the sake of a theory of nature. We wish—ultimately for the sake of ethics—to expand the *ontological locus* of purpose as such from what is apparent at the subjective peak to what is hidden in the breadth of being, without thereafter using the hidden to *explain* anything in what thus contains it but to us shows a visage with whose traits alone we can cognitively deal. Here the following reflection will suffice.

As subjectivity is in some sense a surface phenomenon of nature—the visible tip of a much larger iceberg—it speaks for the silent interior beneath. Or: The fruit betrays something of the root and the stem out of which it grew. Because subjectivity displays efficacious purpose, indeed wholly lives in it, we must concede to the silent interior which finds language only through it, that is, to matter, that it harbors purpose or its analogue within itself in nonsubjective form.

But is it meaningful to speak of "purpose" which is not subjective, that is, mental? And would purpose in matter not clash with the causal explanations of physics? To answer the second question first: It is simply not true that an "Aristotelian" understanding of being contradicts the modern causal explanation of nature or is inconsistent with it, let alone that it has been refuted by it. The objection was against its use in causal explanation, in which the final causes are redundant or question-begging: an otiose assumption at best, and a dangerous one at worst, because invoking them entices the ignorance of true causes into the refuge of supposed knowledge, which excuses from further search (a case of what Spinoza called the *asylum ignorantiae*). As a policy of method, this is entirely correct, and is continuously confirmed by the sole sufficiency of "efficient causes"—that is, of mere quantities of force under constancy laws—in all *individual* explanations, when carried far enough. Whether this sufficiency extends from the individual parts to the explanation of

the whole is moot. It is actually demonstrated only for the artificial simplifications of the experiment or for the most extreme simplifications (the
astronomical) of nature itself.[7] Expectation as well as duty of the explorer
of nature go beyond this proven ground *ad infinitum,* and we need venture
no prediction concerning the success of this never-ending process of re-
ductive analysis. Anyway, explaining nature and comprehending it are
not the same thing. Let us recall that our wish is not to explain nature by
alleged purposes, but rather to interpret the proven occurrence of purposes in it (which is thus not at variance with it) as it bears on the concept
of nature. In trying for a concept of it that accommodates their presence,
we leave entirely open the manner in which a generalized, unconscious
"purposiveness" of nature may assert itself in its deterministic causal
mechanism, not so much against as through it: just as natural science
must leave open how tight or loose, how precise or blurred, the causal
web really is at the undermost base of things (beneath a certain threshold
of size). For natural science it is enough that in the *measurable* regions
the quantitative-deterministic accounting always tallies, that is, that its
equations each time stand the test of event and its method is rebuffed by
none. And that is quite compatible with an underlying teleology of the
whole train of events. We are thus actually saying no more than that
natural science doesn't tell us everything about nature—whereof its avowed
inability to ever account, from its premises, for consciousness, nay, for
the most elementary case of feeling (thus for the best-evidenced phenomenon in the universe!) is the most telling and conclusive evidence. It is
an essential, not a provisional inability, and a paradoxical side-effect of
it is that science itself, as an occurrence within the universe which it
undertakes to explain, is forever excluded from what *it* can explain. Its
own existence is indeed its own best corrective.

d) Purpose beyond subjectivity: what sense does it make?   There remains
the question of what sense it makes to speak of a "purpose" which is
not entertained by a subject in his subjectivity, a purpose not in anyone's
"thought." Can one meaningfully speak of nonmental purpose? For it
would, of course, be the height of folly to assert immanence of purpose
or goal in the digestive organ, in the cells of the body, in primitive organisms, also in the process of evolution, if this term included *mentality*
of any sort—not to speak of conscious intention complete with an idea
of the goal. With the imputation of this folly, the polemic on behalf of
science against the *concept* of teleology (as distinct from the polemic
against its explanatory role) has always made things too easy for itself.
Admittedly we know of goals in general first and foremost through what
we *know,* that is, are aware of *in ourselves*—hence, of conscious goals
only (even, strictly speaking, everybody only of his own). To say this is
a truism, since purpose is not a thing visible from without. But even in

the brightness of our intensified mentality we know of the more and the less conscious, of degrees of awareness; and to speak of a dark urge, even of unconscious wishes and struggles in ourselves, is not considered at all nonsensical. When hence we descend, from man down along the animal tree, the principle of continuity requires us to concede an endless shading, in which "representational" subjectivity surely disappears somewhere (presumably in forms with no specific sense organs yet), but sensitivity and appetition as such probably nowhere.[8] Even here, to be sure, we are still dealing with "subjectivity," but with one already so diffuse that the concept of an individual, focused subject gradually ceases to apply, and somewhere the series trails off into the complete absence of any such subject. Therefore also into an absence of aim and urge? Not necessarily. On the contrary: in the reverse direction, ascending from the bottom upward, it would be incomprehensible that subjective striving in its particularization should have emerged without striving whatever within the emergence itself. Something already kindred must have carried it upward out of the darkness into the greater light.

Admittedly, a "psychic" aspect always adheres to striving as such. And why not? "Psyche" and "selfhood" are not identical, and the first may in a generalized form be an appurtenance of all matter, or of all material aggregates of certain forms of order, long before it attains individuation and hence the horizon of *selfhood* in highly organized, metabolizing aggregates that set themselves off from and actively trade with the environment—"independent" organisms. Should anyone insist on ascribing "subjectivity" already to the diffuse psychic realm *beneath* this, we need not quibble over the semantic issue: it would then either be a subjectivity without a "subject," or else "Nature" could be called its impersonal subject—an unconscious total subject, whatever this might mean, not an individual subject distinguishing itself from others. But diffuse appetitive striving seems to me not to need such a hypostatization at all, and, in light of the distribution of lifeless matter in the universe, I would sooner believe in subjectivity without a subject, that is, in a scattering of germinal appetitive inwardness through myriads of individual particles, than in its initial unity in a metaphysical, all-embracing subject. (In other words, pantheism is not a necessary complement of panpsychism.) Distinct "units" of ordered groupings (patterns) of the manifold, whether organic or inorganic, would then already be an advanced result, a crystallization as it were, of that scattered "aiming," and it would carry with it difference from the surroundings, or individuation. But such speculations go far beyond what we need here.

At any rate, we repeat, just as manifest subjectivity (which is always particular as well) is something of an upstart surface-phenomenon of nature, so too is it rooted in that nature and stays in continuity of essence with it; and that continuity makes both participate in "purpose." On the strength of the evidence of life (which we of its stock in whom it has come to know itself should be the last to deny), we say therefore that purpose in general is indigenous to nature. And we can say something more: that in bringing forth life, nature evinces at least *one* determinate purpose— life itself. This perhaps means nothing else than the setting free of "pur- pose" in general for definite, also subjectively pursued and enjoyed pur- poses. We refrain from saying that life is "the" purpose or the main purpose of nature—these things being beyond our guessing; it is enough to say "a purpose." If, however (by a not unreasonable conjecture), "purposive being" itself were the basic purpose, in a manner of speaking the purpose of all purposes, then indeed life, in which purpose is set free, would be an eminent form of bringing *this* purpose to fruition.

e) Willing, occasion, and the channeling of causality   One more remark to clarify the sort of "volition" here ascribed to nature. It is a will to go beyond itself, but need not be linked with "knowledge," certainly not with knowing and figuring to itself the goal in advance. But discrimination it must possess—so that at encountering the physically favorable config- uration, causality is not indifferent to its invitation, but gives it selective preference and rushes into the offered opening, thereafter to carve its bed through further opportunities as they offer themselves.[9] To what extent "aiming" can itself bring about such opportunities—that is, where the direct power of interest begins, as distinct from the mere, indirect ex- ploiting of good fortune—is another question. The circumspect assump- tion is that a goal orientation is there and avails itself of its opportunities. Conversely, it may be also that new opportunities give rise to new, pre- viously unknown goals, and for this reason it would perhaps be better (in any case, still more circumspect) to speak of goal disposition rather than of goal orientation. (How many are the human enterprises which meander through their course in this manner!) But such "suggestions" of goals by chance occasion and the turns in direction they cause would probably apply more to single segments of the course than to its overall direction; and even the occurrence of the suggestive occasions could have been helped along already by the earlier goal-orientation—which then might indeed be surprised by the possibilities opening up in its wake. On this we can only speculate, not settle anything—especially regarding the "first" opportunity, with which "life" began. But even assuming the first begin- ning, the forming of organic macromolecules, to have been mere accident, not the fulfillment of a preceding tendency (to me, a most unlikely as- sumption)—from there on, certainly, tendency becomes ever more ap- parent: and I mean not only a tendency for progressive evolution (which can rest as long as it pleases) but above all the tendency *to be,* ceaselessly at work *in* each of its creations. I have elsewhere attempted to show how already in the "simplest" true *organism*—existing by way of metabolism, and thereby self-dependent and other-dependent at once—the horizons of selfhood, world and time, under the imperious alternative of being or nonbeing, are silhouetted in a premental form.[10]

In conclusion, then, to answer the title question of IV.2 (p. 66) and its more pointed framing in IV.3.*d* (p. 72): it is meaningful, and not just a metaphor borrowed from our subjectivity, to speak of the immanent, if entirely unconscious and involuntary, *purpose* of digestion and its apparatus in the totality of the living body, and to speak of life as the end-purpose of just this body. It makes sense and has greater probability on its side[11] than its opposite, to speak of a "laboring" in nature and to say that nature along its tortuous paths is laboring toward something, or this something multifariously struggles in it to come to light. Even if this only began with the "accident" of life, that would be enough, for purpose is thereby extended beyond all consciousness, human and animal, into the physical world as an innate principle of it. How far its sway in living things reaches down into more elementary forms of being can remain an open question. A room and a readiness for it must be credited to the being of nature as such.

V. From Reality to Validity: From the Problem of Purpose to the Problem of Value

Does this expansion of "purpose" across reality release it from the bane of "subjectivity"? Does universality of occurrence confer validity? Have we thus won something for ethics, which says values *ought* to be purposes if and because they have objective validity, by demonstrating that purpose is *de facto* present in nature, indeed is embedded in the nature of things? Can nature legitimate purposes by having them? It is the old question of whether being can in any way ground obligation. This is for the next chapter to discuss, which is devoted to the status of values. But by way of transition we close this chapter with some observations concerning the relation of universality and validity, as well as concerning the relation of "mere" subjectivity to one that permeates nature.

1. *Universality and Legitimacy*

All men, so it is said, desire to be happy; and it is said not as a statistical observation but with the tacit corollary that this lies in man's very nature, that is, as a statement of essence. Now, the universality of the goal of happiness so stipulated is by itself a mere fact, even if a necessary fact, of nature: one is not, so it seems, bound thereby to adopt its pursuit for oneself and approve it in others; one may roundly despise or condemn it. But it must be conceded that it is not arbitrarily chosen, and the fact that it is so universally implanted in our nature creates at least a strong presumption for its pursuit to be justified: it can be taken to indicate, if not a duty, at any rate a *right* to this goal, "happiness"—thus to indicate that we are at least permitted, perhaps even obliged, to pursue it (subject to certain moral conditions). This, however, would in turn lead beyond one's own right to the duty to respect this same right in others, thus not to hinder it, perhaps even to promote it. And from this interest of others, which is entitled to my respect, there could even follow indirectly (if it does not exist directly) a duty for me to promote my own happiness as well, whose stunting would disturb the general happiness . . . If a reasoning on these lines is allowed, then the *de facto* universality of the aspiration for happiness, which engenders a presumption of its justness, does after all contribute to its *de jure* moral standing. The backing of nature, evinced in the universality, would not be altogether negligible for its backing by ethics. We speak of a "presumption only," as ours here is a provisional deliberation and not as yet a serious philosophical proof. For whether such things as right and duty, the licit and the illicit, do exist at all is left entirely open in the above argument, which simply presupposes them. But *if* they exist, then the vote of nature may become an important indicator for determining them and perhaps even an authority for sanctioning them. This question must wait.

Let us look, however, in our example (which was chosen from convention and for illustration only), at the distinction between the universal and the particular. The *question* of right and value normally arises in relation to the particular goals and wills of individual subjects and their possible conflicts, while both right and value are readily granted to what they all are agreed about; and it is there, in regard to particular endeavors, that we are most inclined to speak of "mere subjectivity" of purposes and valuations, and there alone where we speak of arbitrariness which must first justify itself. Now, this we can say with certainty of a "subjectivity" of nature, that it is neither particular nor arbitrary, and that over against our private desirings and opinings it has all the advantages of the whole over the parts, of the abiding over the fleeting, of the majestic over the puny.

2. We Are Free to Reject the Vote of Nature

Nevertheless, we can oppose its vote, that is, the partisanship of its purposes, even though in so doing we avail ourselves of *one* of these purposes—namely, *freedom*. It is the prerogative of human freedom to be able to say no to the world. That the world has values indeed follows directly from its having purposes (and this having been shown to be the case, there can in this sense no longer be talk of a "value-free" nature), but I need not *share* its "value-judgments" and can even decide, with Goethe's Mephisto, "Whatever is, deserves to perish. Hence it were better, nothing came to be." It must also be conceded that the sheer superiority of nature in size, duration, and power, even in the splendor of its creations, does not by itself confer an *authority* on its value options that would make them binding for me—just as little, one might demagogically add, as among men the numerical superiority of mass opinion over the minority establishes what is true.

Yet surely (for a first rejoinder) it is a very different matter, in quality and not just quantity, to oppose the deepest-rooted, oldest, and most constant vote of nature, of which we are ourselves, than to oppose the fickle, short-lived, and "superfical" opining of men. More to the point, I can *legitimately* dissent from nature only if I can appeal to a tribunal outside it—that is, to a transcendence which I believe to possess the authority I deny to the former: hence, only by embracing some kind of dualism. (A dissent based upon mere taste or mood would be frivolous.) This dualism would have to work with the theology of a God who is either not responsible for the world at all, or was crossed in its creation by an opposing principle, or for some higher purpose of his own created it flawed himself; and hence with the concept of an accordingly bad (not just indifferent) world. Moreover, and in any of these cases, it would require the doctrine of a transcendent soul as performing that act of dissent. In short, a wholesale, cosmic dissent, to be intellectually respectable, would call for a "gnostic" theory of being—probably the last thing anyone involved today in the quarrel about a "meaningful" or "meaningless" world would be willing to entertain. Under monistic terms, however, legitimate dissent is possible only in particulars, not concerning the whole.

3. *Rebuttal of the Right to Negate Proves No Duty to Affirm*

Still, the impossibility of legitimate *negation* does not suffice to legitimize the object thereof, that is, to entitle it to *affirmation*. (At least fictively, a standpoint of summary neutrality is conceivable—just as toward an indifferent nature which indeed leaves nothing but personal taste or mood to respond to particular phenomena.) Not even the fact that in dissociating myself from the whole, which is an exercise of the freedom given me, I am in fact endorsing Nature in this eminent one of her value-decisions (or whims), obliges me to say "yes" to her, as logic might seem to dictate. Logic can only incriminate the inconsistency of my sweeping "no" with the particular "yes" implicit in pronouncing it, namely, with the self-affirmation of the interest-charged moment (like that of every use of life except suicide); but it cannot, in the name of consistency, turn this deviant, *de facto* affirmation into an authentic, *de jure* affirmation by ourselves. For this, that is, for real, obligatory affirmation, the concept of the *good* is needed, which is not identical with the concept of value, or if you will, which signifies the distinction between objective and subjective status of value (or at its briefest: between value in itself and valuation by someone). And it is the relationship between goodness and being (*bonum* and *esse*) with whose clarification a theory of value can hope to ground a possibly binding force of values—namely (to utter the immodest aim), by grounding the good in being. Only hence can it be argued that Nature, in adhering to values, also has the authority to sanction them and may demand their acknowledgment by us and by every knowing will dwelling in her midst.

Thus our showing up to now that Nature harbors values because it harbors ends and is thus anything but value-free, has not yet answered the question of whether we are at pleasure or duty bound to join in her "value decisions": whether, to put it paradoxically, the values undeniably entertained by and for herself are indeed valuable (even whether *having* values as such is valuable!)—in which case alone assenting to them would be a duty. Not the theory of ends—which showed values to be present in nature as the objects of purpose—can answer that question, but the theory of value must do so, to which we now turn. But showing the immanence of purpose in nature made asking this question now possible; and it will turn out that, with the gaining of this premise, the decisive battle for ethical theory has already been won.

4
The Good, the "Ought," and Being: A Theory of Responsibility

I. Being and Ought-to-Be

To ground the "good" or "value" in being is to bridge the alleged chasm between "is" and "ought." For the good or valuable, when it is this of itself and not just by grace of someone's desiring, needing, or choosing, is by its very concept a thing whose being possible entails the demand for its being or becoming actual and thus turns into an "ought" when a will is present which can hear the demand and translate it into action. Thus we say that a "command" can issue not only from a commanding will, for instance, of a personal God, but also from the immanent claim of a good-in-itself to its realization. If, however, the "good" or "value" is indeed something by itself, then it belongs to the stock of being in general (not therefore necessarily to the actuality of what happens to exist at a given time), and, in that case, axiology becomes a part of ontology. How does this relate to our earlier findings about the essence of nature?

1. *"Good" or "Bad" as Relative to an End*

Nature, by entertaining ends, or having aims, as we now assume her to do, also posits values. For with any *de facto* pursued end (whatever made it into this), attainment of it becomes a good, and frustration of it, an evil; and with this distinction the attributability of value begins. But from *within* the predecided goal commitment, where henceforth only success or failure counts, no judgment upon the goodness of the goal itself is possible, and thus beyond the factual interest, no obligation can be derived from it. Insofar, then, as ends, including our own, are actually at play within nature, they seem to enjoy no other dignity than that of mere facts and would then have to be measured not by worth but only by their motivating strength and perhaps by the pleasure yield of their achievement (or the pain of their denial). We could then only say that under their spell there is a better and a worse, but not that through them a good-in-itself demands our affirmation. Is there then any sense in saying that anything *ought* to be, no matter whether or not it promotes its coming-to-be on its own by influencing desire, instinct, or will? We said that a "good-in-itself" would be such a thing. But up to now good or evil has only shown itself as the correlate of an already *existing* goal-orientation, to which it is left to exercise such power over the will as is *ex post facto* attested by the will's "decisions"—the result of just that power. The implanted purpose looks after itself and requires no "ought"; nor could it provide a ground for it in itself. At best it might use the fiction of an "ought" as a means of enhancing its power.

2. Purposiveness as Good-in-Itself

But what is true of the particular purpose—namely, that the fact of it comes first, and the validity of "good" and "bad" relative to it comes second, determined by the first (*de facto*), but not legitimized by it (*de jure*)—is that also true of "purposiveness" itself, as an *ontological* characteristic of an entity? Here, it seems to me, matters are different. We can regard the mere *capacity* to *have* any purposes at all as a good-in-itself, of which we grasp with intuitive certainty that it is infinitely superior to any purposelessness of being. I am not certain whether this is an analytic or a synthetic statement, but whatever of self-evidence it possesses, there is plainly no going back behind it for something more basic to underpin it. One can only oppose it with the doctrine of nirvana, which denies the value of having purpose, but then again affirms the value of liberation from it, making this its purpose. Since indifference is clearly not possible here (what is denied becomes a negative value), at least he who does not embrace the paradox of a purpose-denying purpose must concur in the proposition that purpose as such is its own accreditation within being, and must postulate this as an *ontological axiom*.[1] Now, it follows already analytically from the *formal* concept of the good-in-itself that, whenever this first, self-validating good happens, in any of its individuations, to come under the custody of a will, it addresses an "ought" to this will. And the *content* of this first good (and at the same time the recognition of its rooting in reality) is nothing else than what is affirmed in that first axiomatic intuition: the superiority of purpose as such over purposelessness. Even if our granting it the dignity of an axiom (at first an act of pure theory) is a matter of ultimate metaphysical choice which can give no further account of itself (perhaps as little as can the self-choice of being which it underwrites), still it commands an evidential intuition of its own, which can be articulated roughly as follows.

3. *The Self-Affirmation of Being in Purpose*

In purposiveness as such, whose reality and efficacy in the world we take as settled by the preceding discourse (Chap. 3), we can see a fundamental self-affirmation of being, which posits it *absolutely* as the better over against nonbeing. In every purpose, being declares itself for itself and against nothingness. Against this verdict of being there is no counterverdict, for even saying "no" to being betrays an interest and a purpose. Hence, the mere fact that being is not indifferent toward itself makes its difference from nonbeing the basic value of all values, the first "yes" in general. This difference rests not so much in the distinction of a something from nothingness (which with a value-indifferent something would merely be the—in turn indifferent—distinction between two matters of indifference), but rather in the distinction of goal-interest as such from indifference as such, of which we could regard nothingness to be the absolute form. An indifferent being would be only a less perfect form (because afflicted with the blemish of senselessness) of nothingness and not really thinkable. That being is concerned with something, at least with itself, is the first thing we can learn about it from the presence of purpose within it. Then the next value, deriving from the basic value of being as such and enhancing its difference from nonbeing, would be the maximization of purposiveness, that is, the growing wealth of goals striven for and thus of possible good or evil. The more manifold the purpose, the greater the difference; the more intensive it is, the more emphatic the affirmation and, at the same time, its justification. In it, being makes itself worth its own effort.

4. *The "Yes" of Life: Emphatic as the "No" to Nonbeing*

In organic life, nature has made its interest manifest and progressively satisfies it, at the rising cost of concomitant frustration and extinction, in the staggering variety of life's forms, each of which is a mode of being and striving. The price is necessary, since each purpose can be realized only at the cost of other purposes. The generic multiplicity itself represents such a selection, of which it is impossible to say whether it was always the "best," but whose preservation is certainly a good compared to the alternative of annihilation or impoverishment. But more still than in the extensiveness of the generic spectrum, the interest manifests itself in the intensity of goal-striving of the living creatures themselves, in which the natural purpose becomes increasingly subjective, that is, increasingly the individual executants' very own. In this sense, every feeling and striving being is not only an end of nature but also an end-in-itself, namely, its own end.[2] And precisely here, the self-affirmation of being becomes emphatic in the opposition of life to death. Life is the explicit confrontation of being with not-being. For in its constitutional neediness, given with the necessity of metabolism, which can be denied fulfillment, it contains within itself the possibility of not-being as its ever-present antithesis—as threat. Its mode of being is preservation through doing. The "yes" of all striving is here sharpened by the active "no" to not-being. Through the negated not-being, being becomes a positive concern, that is, a constant choosing of itself. Life as such, in the inherently co-present danger of not-being, is an expression of this choice. Thus it is only an apparent paradox that it should be *death,* that is, the being liable to die (being "mortal") and being so at every moment, and its equally ceaseless deferment every moment by the *act* of self-preservation, which sets the seal upon the self-affirmation of being: in this contrapuntal pairing, the self-affirmation of being turns into single efforts of individual beings.

5. *The Obligating Force of the Ontological "Yes" upon Man*

This blindly self-enacting "yes" gains obligating force in the seeing freedom of man, who as the supreme outcome of nature's purposive labor is no longer its automatic executor but, with the power obtained from knowledge, can become its destroyer as well. He must adopt the "yes" into his will and impose the "no" to not-being on his power. But precisely this transition from willing to obligation is the critical point of moral theory at which attempts at laying a foundation for it come so easily to grief. Why does now, in man, that become a duty which hitherto "being" itself took care of through all individual willings? Why this standing out of man from nature, according to which he is supposed to come to the aid of her governance with norms and therefore to impose limits on his own unique, natural inheritance, arbitrary will? Would not the fullest exercise of just this arbitrariness be the fulfillment of the natural purpose which brought it forth—wherever it might lead? Precisely this would be the value in itself toward which the movement of being had striven, and this its vote, which could demand assent but does not need it at all.

6. *The Problematic Nature of an "Ought" Distinct from the Will*

Granting then that "purposiveness" is in itself the primary good and as such has, in abstract terms, a "claim" to realization, it already involves a *willing* of ends and through them, as conditions of its own continuance, the willing of itself—the basic end. Thus natural purposiveness itself looks after the fulfillment of its claim to being, which is consequently in good hands. Simply stated, self-preservation need not be commanded and needs no persuasion beside the pleasure associated to it. Its willing, with its "yes" and "no," is always there from the beginning and takes care of its business—better or worse as the case may be, but always to the measure of capacity. Thus even if "ought to will" were a meaningful concept, it would be superfluous here, as would be the (actually meaningful) concept of "ought to do," since the willing, once present, carries its doing automatically with it. Where, however, there is a *choice* between a better and a worse (i.e., a more and a less effectual), as in the case of men, there one can indeed speak, in the name of the willed end, of an "ought" for choosing the better *way*—thus (with Kant) of a "hypothetical imperative" of prudence, which concerns the means and not the end itself. But as important as this imperative can become in the jumble of human affairs—it has little to do with the unconditional imperative of morality. This latter must extend as well to purposes, indeed to them above all. (That it is its own purpose, as Kant in the last analysis thought, is an untenable construction—see below.) Nor does it help to speak of the superiority of "higher" over "lower" grounds for choice, as long as this distinction is not already ethically defined and something like a duty to the higher end is not demonstrated. One may with good reason consider the painting of the Sistine Chapel's ceiling a higher purpose than the quenching of a gnawing hunger, but one should ask Heine's "Migratory Rats"[3] whether for them a law of action follows hence. From the objective validity of value or "the good," which we here already presume, together with its abstract "claim," it is yet a further step to the *task* which is posed to action here and now as, at this moment, *mine:* the step from the timeless into time. Behind this step lurks the suspicion that perhaps even the odd self-deception of morality, be it the most ascetic, with all its "tasks" and "renunciations," is but a camouflaged form of satisfying the primal urge (for instance, "the will to power," "the pleasure principle")—hence that all self-imposed, putative "ought" is in truth only a disguise for the will, its seduction by a more effective bait than that of base pleasure. In this case, "value" or "the good" would not have the authority of command but the strength of causes—of final causes, to be sure, but a causal strength no less. Then all willing as such, and as part of the immanent teleology of being in general, would be *eo ipso* justified (to be appraised perhaps according to its intensity, but not according to its goal), and trying for a doctrine of duties would be a vain endeavor. Even the resounding "yes" of Nietzsche's *amor fati*, the empty willing again of everything willed and done before, would add nothing to this blank. We must examine once again the meaning of "value" and of "good."

7. *"Value" and "Good"*

a) Linguistically, "the good" as compared with "value" has the greater dignity of objective status. We are inclined to understand it as something independent of our wishing and thinking. "Value" on the other hand is easily tied to the question "For whom?" and "How much?" The word stems from the sphere of appraisal and exchange. It thus designates at first only a measure of willing, namely, of the will to spend, and not of an honor due. I set myself something as an end because it is worth something to me; or, it is worth something to me because it is already set as an end to my indigent nature, prior to all choosing, by one of its congenital needs: in acting on it, insofar as this was my choice in the competition among ends, I make the specific natural end my actual own for the moment.[4] Thus *every* purpose I set myself is by this alone identified as a "value," that is to say, as worth to me now the effort of pursuing it (plus the forsaking of those I can consequently not pursue). The exchange value of the effort—its "recompense"—is here pleasure, including its most refined varieties. If the attained end disappoints me in this and leads me to the conclusion that it was not worth the effort after all, then in the future, too, my better informed desire will only consult itself regarding a more rewarding choice of ends, but not these ends themselves regarding their claim upon my choice. Even the revised judgment, although better informed and thus perhaps more successful, need be no less subjective and hence no more binding than the original one.

b) Nevertheless, we do not let ourselves be dissuaded from distinguishing between worthy and unworthy ends, and this independently of whether or not desire will get a good bargain out of them. We postulate with this distinction that what *is* worth my effort does not of itself coincide with what just *appeals to me* as worth my effort. However, what is truly worth my effort *should* also become *to me* a matter worth it and therefore *made* by me into a purpose. Now, "truly worth the effort" must mean that the object of the effort is *good,* independent of the verdict of my inclinations. Precisely this makes it the source of an "ought," with which it addresses the subject in the situation where the realization or preservation of *this* good by *this* subject is a concrete issue. No voluntaristic or appetitive theory which defines the good as what is desired does justice to this primordial phenomenon of *demanding*. If the good is a mere creature of the will, it lacks the authority to bind the will. Instead of determining its choice, it is subject to it and is one time this, another time that. Only its foundation in Being places it over against the will. The independent good demands that it become purpose. It can not compel the free will to make it its purpose, but it can extort from it the recognition that this would be its duty. If not in obeying, this recognition manifests itself in the feeling of guilt: we failed to give the good its due.

8. *Doing Good and the Being of the Agent: Predominance*
   of the "Object"
   a) However, just as little as we are willing to part with the distinction between desire and obligation will our feeling let go of the certainty that doing the good for its own sake also in some sense benefits the agent, and this regardless of the success of his action. Whether he is allowed to enjoy the achieved good himself, or not; whether he lives to see it achieved, or not, even should he see his action fail—his moral being has gained with the obedient acceptance of the call of duty. Yet not *this* must have

been the good he willed. The secret or paradox of morality is that the self forgets itself over the pursuit of the object, so that a higher self (which indeed is also a good in itself) might come into being. To be sure, I am allowed to say, "I wish to be able to look myself in the eye (or: pass God's scrutiny)," but just this will be possible only if my concern was with the object and not with myself: the latter must not itself become the object, and the deed's object only the occasion for it. The good man is not he who made himself good but rather he who did the good for its own sake. But the good is the "cause" at issue out there in the world, indeed the cause of the world. Morality can never have itself for its goal.

b) Consequently, it is not the form but the content of action which takes priority. In this sense morality is "selfless," although it may sometimes also have a condition of the self for its object—namely, a condition conforming to duty, as part of the "cause" of the world (and without selflessness *per se* being moral). Not duty itself is the object; not the moral law motivates moral action, but the appeal of a possible good-in-itself in the world, which confronts my will and demands to be heard—*in accordance with* the moral law. To grant that appeal a hearing *is* precisely what the moral law commands: this law is nothing but the general enjoinder of the call of all action-dependent "goods" and of their situation-determined *right* to just *my* action. It makes my duty what insight has shown to be, of itself, worthy of being and in need of my acting. For that enjoinder to reach and affect me, so that it can move the will, I must be receptive for appeals of this kind. Our emotional side must come into play. And it is indeed of the essence of our moral nature that the appeal, as insight transmits it, finds an answer in our feeling. It is the feeling of responsibility.

c) A theory of responsibility, as any ethical theory, must deal both with the rational ground of obligation, that is, the validating principle behind the claim to a binding "ought," and with the psychological ground of its moving the will, that is, of an agent's letting it determine his course of action. This is to say that ethics has an objective side and a subjective side, the one having to do with reason, the other with emotion. Sometimes the one, sometimes the other has been more in the center of ethical theory, and traditionally the problem of validation, that is, the objective side, has posed the greater challenge to philosophers. But the two sides are mutually complementary and both are integral to ethics itself. Without our being, at least by disposition, responsive to the call of duty in terms of feeling, the most cogent demonstration of its right, even when compelling theoretical assent, would be powerless to make it a motivating force. Conversely, without some credentials of its right, our *de facto* responsiveness to appeals of this kind would remain at the mercy of fortuitous predilections (variously preconditioned themselves), and the options made by it would lack justification. This, to be sure, would still leave room for moral behavior from a naively good will whose direct self-certainty asks for no further validation—and, indeed, may not need it in those favored cases where the promptings of the heart are "naturally" in unison with the biddings of the law. A subjectivity so graced (and who will exclude the possibility of it?) could act all by itself. No similar sufficiency can ever be enjoyed by the objective side: its imperative, evident as its truth may be, cannot become operative at all unless it meets with a sensitivity to the like of it. This sheer fact of feeling, presumably a universal potential of human experience, is thus the cardinal datum of the moral life and, as such, implied in the "ought" itself. It is indeed of the very meaning of the normative principle that its call is addressed to recipients so constituted that they are by nature receptive to it (which does not, of course, already insure its being heeded). One may well say that there would be no "thou shalt" if there were no one to hear it and on his own attuned to its message, even straining toward its voice. This is the same as saying that men already *are* potentially "moral beings" by possessing that affectibility, and only thereby can they also be immoral. (Those who are by nature deaf to this voice can be neither moral nor immoral.) But it is equally true that the moral sentiment itself demands its authorization from beyond itself, and this not merely in defense against challenges from without (including those from rival motives in oneself) but from an inner need of that very sentiment to be in its own eyes more than a mere impulse. Not the validity, to be sure, only the efficacy of the moral command depends on the subjective condition, which is its premise and its object at the same time, solicited, appealed to, claimed with success or in vain. In any case, the gap between abstract validation and concrete motivation must be bridged by the arc of sentiment, which alone can sway the will. The phenomenon of morality rests *a priori* on this correlation, even though one of its members is only *a posteriori* given as a fact of our existence: the subjective presence of our moral interest.[5]

In the order of logic, the validity of obligation would have to come first and the responding sentiment second. But in the order of approach, there is advantage in beginning with the subjective side, as not only the immanent given but also as implied in the transcendent summons directed at it. We can take only the briefest of looks at the emotional aspect of morality in past ethical theory.

9. The role of sentiment in past ethical theory

a) Love of the "highest good"   Moral philosophers have always recognized that feeling must supplement reason so that the objectively good can exert a force on our will; in other words, that morality, which is meant to have command over the emotions, requires an emotion of its own to do so. Among the great ones, it was probably Kant alone who had to wring this from his rigorism as a concession to our sensuous nature, instead of seeing it as an integral aspect of the ethical as such. Explicitly or implicitly, the insight lives in every doctrine of virtue, however differently the feeling in question may have been determined. Jewish "fear of the Lord," Platonic "eros," Aristotelian "eudaimonia," Christian "charity," Spinoza's *"amor dei intellectualis,"* Hutcheson's "benevolence," Kant's "reverence," Kierkegaard's "interest," Nietzsche's "lust of the will" (and so on) are modes of defining this affective element in ethics. We cannot discuss them here, but we observe that the feeling of responsibility is not among them. This absence must be explained later in defense of our choice. We also observe that most, though not all, of the feelings named were of the kind inspired by and directed at an *object* of supreme worth, a "highest good," which often carried the ontological connotation (a corollary to the idea of perfection) that this must be something timeless, confronting our mortality with the lure of eternity. The aim of ethical striving is, then, to emulate this object in our own condition and to this extent to "appropriate" it and also to aid its appropriation by others: in general, to create a place for it within the world of time. The imperishable invites participation by the perishable and elicits in it the *desire* thereto.

By contrast the object of *responsibility* is emphatically the perishable *qua* perishable. Yet in spite of this condition which it shares with myself, it is more unsharably an "other" to me than any of the transcendent objects of classical ethics; "other" not as the surpassing better, but as nothing-but-itself in its own right, and without *this* otherness being meant to be bridged by a qualitative assimilation on my part or on its part. Precisely this otherness takes possession of my responsibility, and no appropriation is intended here. Yet just this far from "perfect" object, entirely contingent in its facticity, perceived precisely in its perishability, indigence, and insecurity, must have the power to move me through its sheer existence (not through special qualities) to place my person at its service, free of all appetite for appropriation. And it evidently has this power, or else there would be no *feeling* of responsibility for such an existence. But there is the fact of such a feeling, given in experience and no less real than the appetitive feelings of the *summum bonum* aspiration. Of this, we have to speak later. Here we note what is nonetheless common between the two contrasting types: that the committing force issues from the claim of an object, and the commitment is to that object, whether eternal or temporal. In both cases, something is to be brought about in the order of things.

b) Action for the sake of the action Over against these object-inspired and object-committed ethical stances, in which the content of the aim reigns supreme, stand the objectless kinds, in which the form or spirit of the action itself is the theme of the norm, and the external object, provided by the situation, is more the occasion than the aim for the deed. Not the "what" but the "how" of acting really matters. Existentialism is the modern extreme of this ethics of subjective intention (cf. Nietzsche's "will to will," Sartre's "authentic decision," Heidegger's "resoluteness," etc.), where the worldly issue is not by itself endowed with a claim on us but receives its significance from the choice of our passionate concern. Here the self-committing freedom of the self reigns supreme. Whether this position is tenable and does not hide a surreptitious recognition of a value in the object itself for which the decision opts (for which it therefore *ought* to opt), and whether this is not the true ground for the allegedly groundless choice, need not be discussed here. What matters for ethical theory is the conceptual denial of any immanent order of rank or right among objects and, therefore, of the very idea of objectively valid obligations toward them of which they themselves could be the source.[6]

c) *Kant's "reverence for the law"*   Unique (as in so many other respects) is Kant's position in this quarrel between "material" and "formal," "objective" and "subjective" principles of moral action. While not denying that objects can affect us by their worth, he denies (for the sake of the "autonomy" of reason) that this emotive affection supplies the true motive for moral action; and while stressing the rational objectivity of a universal moral law, he concedes the necessary role of feeling in conforming to it. What is unique is that this feeling is directed not at a material object but at the law itself. It was indeed among the profound insights of Kant, the more telling for coming from the champion of unadulterated autonomy of reason in moral matters, that besides reason there must also be sentiment at work so that the moral law can gain the force to affect our will. According to him, this was a sentiment evoked in us not by an object (which would make the morality heteronomous), but by the *idea* of duty, that is, of the moral law itself; and this sentiment was "reverence" (*Ehrfurcht*). Kant thought: reverence for the law, for the sublimity of the unconditional "thou shalt" that issues from reason. In other words, reason itself becomes the source of an affect and its ultimate object. Not, of course, reason as a cognitive faculty, but reason as a principle of universality, to which the will is enjoined to conform; and this not through the choice of its objects, but through the form of choosing them, that is, through the mode of determining itself with a view to possible universalization. This internal form of willing alone is the content of the "categorical imperative," whose sublime right is said to inspire us with reverence.

But this thought, although not without sublimity itself, leads into absurdity. For the meaning of the categorical imperative is, as all its casuistic applications show, not the setting of ends, but the self-limitation of freedom, through the rule of the consistency of the will, in the pursuit of ends. But if this is the idea of the moral law, then the Kantian formula amounts in the end to "self-limitation of freedom out of reverence for the idea of the self-limitation of freedom"—which is obviously vacuous. Or, since the self-limitation is to proceed by the criterion of the act's permitting "universalization" (without self-contradiction), one could also say, "universalization of the particular will out of reverence for the idea of universality," which is only a little better. For, admittedly, wider generality is a virtue in the case of theoretical propositions in a system of truth, and their validity for every other mind goes without saying; but in the case of individual decisions on acting, the possible accompaniment by a certainty that every reasonable being must agree with them on account of the generality of their principle, may well be a welcome support (perhaps even a sign of their correctness) but surely cannot be the first ground for my choice, and most surely not the source of the *feeling*—be it reverence or something else—which here and now animates my compact with the matter. Only the object itself, and no idea of its generality, can evoke this feeling, and it does so precisely by its unique validity as itself. This may in turn stand under more comprehensive principles, but these would be ontological, and if such principles affect the sentiment it is through their *content* and not their degree of generality. (Every attempt to understand the moral law as its own purpose must similarly fail as this greatest example.)

In truth, it must be added, Kant's moral insight was greater than what the logic of the system dictated. The peculiar vacuity into which the purely formal "categorical imperative" leads with its criterion of noncontradictory "universalizability" of the maxim of the will, has been often noted.[7] But Kant himself has redeemed the mere formality of his categorical imperative by a "material" principle of conduct, ostensibly inferred from it but, in fact, added to it: respect for the dignity of persons as "ends in themselves." To this, the charge of vacuity surely does not apply! But the unconditional self-worth of reasoning subjects here posited follows from no formal principle, but must convince the value sense of the judging observer who, with the "eyes of the mind," *beholds* what a freely acting self in a world of necessity means.

d) The standpoint of the following investigation Our counterposition, which underlies the following reflections on responsibility, can be stated quite simply: what matters are things rather than states of my will. By engaging the will, the things become ends. Ends may sometimes be sublime—by *what* they are, and even certain acts or lives may be so; but not the formal rule of the will whose observance is for any chosen end, or act, the *condition* of being a moral one, or, more precisely, of not being an immoral one. The *law* as such can be neither the cause nor the object of reverence; but *Being* (or instances of it), disclosed to a sight not blocked by selfishness or dimmed by dullness, may well instill reverence—and can with this affection of our feeling come to the aid of the, otherwise powerless, moral law which bids us to honor the instrinsic claim of Being. To be "heteronomous" in this way, that is, to let oneself be moved by the just appeal of entities, need not be feared or disclaimed in the cause of pure principle. Yet not even "reverence" is enough, for this emotional affirmation of the perceived dignity of the object, however vivid, can remain entirely passive. Only the added *feeling of responsibility,* which binds this subject to this object, will make us act on its behalf. We contend that it is this feeling, more than any other, which may generate a willingness to sustain the object's claim to existence by our action. Finally, let us remember that the care of progeny (see Chap. 2, p. 39 f.), so spontaneous that it needs no invoking the moral law, is the primordial human case of the coincidence of objective responsibility and the subjective feeling of the same. Through it, nature has educated us in advance and prepared our feeling for all the other kinds of responsibility not so buttressed by instinct. Let us then turn to this phenomenon, "responsibility," about which ethical theory has on the whole been so silent.

II. Theory of Responsibility: First Distinctions

The first and most general condition of responsibility is causal power, that is, that acting makes an impact on the world; the second, that such acting is under the agent's control; and third, that he can foresee its consequences to some extent. Under these necessary conditions, there can be "responsibility," but in two widely differing senses: (*a*) responsibility as being accountable "for" one's deeds, whatever they are; and (*b*) responsibility "for" particular objects that commits an agent to particular deeds concerning them. (Note the different referent of "for"!) The one is a formal, the other a substantive concept, and we really speak of two different things when we say that someone is responsible for what happened (which is neither praise nor blame), and that someone is a responsible person, that is, honors his responsibilities (which is praise). Some further articulation will make the distinction clearer.

1. *Formal Responsibility: The Causal Attribution of Deeds Done*
a) "He is responsible, because he did it." That means, the doer must answer for his deed: he is held responsible for its consequences and, where the case warrants it, can be made liable for them. This has primarily legal and not moral significance. The damage done must be made good even if the cause was no misdeed, and even if the consequence was neither foreseen nor intended. It suffices that I was the active cause—but then again only in close causal connection of the consequence with the deed, so that the attribution is clear and the consequence does not lose itself in the unforeseeable. The famous missing horseshoe nail does not really

make the journeyman smith responsible for the lost battle and the fall of the kingdom. But the immediate customer, the horse's rider, would probably have a claim for redress against the blacksmith, who while himself blameless is "responsible" for the carelessness of his journeyman. The carelessness is here anyway the only thing which might be called morally culpable, and that to a trivial degree. But the example shows (as does the commonplace one of parents being liable on behalf of their children) that legal responsibility to indemnify can be free of all guilt. The principle of causal attribution is still preserved in the relationship, by virtue of which the superior generally represents in his person the performance of his subordinates (for whose good performance he also earns the praise).

b) Now, the idea of legal compensation was early intertwined with that of punishment, which has a moral meaning and which qualifies the causative deed as morally blameworthy. It is more the deed than the consequence which is punished in the case of a crime, and it is according to the deed that the measure of expiation is fixed. For this, the deed itself must be investigated—design, premeditation, motive, accountability: Was the deed criminal "in itself"? Conspiracy to commit a crime, which was thwarted by timely discovery, is itself a crime and liable to punishment. The penalty exacted here, with which the doer is called to account, does not serve to compensate for the injury or wrong suffered by others but rather to restore the disturbed moral order. Here, then, it is the quality rather than the causality of the deed which is the decisive point to answer for. Nevertheless, at least potential power remains the *conditio sine qua non*. No one is called to account for the impotent imagining of committing the most horrible misdeeds, and if feelings of guilt are part of the syndrome, they are as private as the psychological offense. A deed must have been committed in the world or at least been begun (as in the conspiracy). And it remains true that the successful deed weighs more heavily than the unsuccessful.

c) The distinction, just indicated, between legal and moral responsibility is reflected in the distinction between civil and criminal law, in whose divergent evolutions the initially intermixed concepts of compensation for injury (out of legal liability) and punishment (for guilt) were separated. Yet, both have in common that the "responsibility" refers to deeds done and becomes real through the doer being *made* responsible from without. The *feeling* which in this context may attend the doer himself and with which he inwardly accepts the responsibility (regret, remorse, readiness to make amends, but also proud defiance) is just as retroactive as the objective "must" of answering for his deed; and even anticipating this consequence on the verge of action does not serve as a motive for acting, but rather (if effective) as a selective screening, that is, as a motive for allowing or blocking execution of the deed. Ultimately one has the less to answer for, the less one does, and in the absence of positive duty the avoidance of action can become the counsel of prudence. So understood, "responsibility" does not itself set ends or disallow ends but is the mere formal burden on all causal acting among men, namely, that they can be called to account for it. As the mere fact of accountability independent of the agent's consent, it is thus the precondition of morality but not yet itself morality. Consenting to it, that is, acknowledging one's accountability, is already more than the choiceless fact. To identify with one's deed, for example, by readiness to take the consequences, has indeed some moral quality which may adorn even utter immorality. That is, owning the deed is better than dissembling it. An example is Mozart's Don Giovanni, whose defiant avowal at the end, paying the extreme price rather than repent, lends a certain grandeur to his misdeeds. But the example also shows that the affirmation of formal responsibility is not a sufficient principle of morality and in its pure formality cannot supply the affective (emotive) element for ethical theory—which is concerned first and last with the presenting, accrediting, and motivating of positive ends directed toward the *bonum humanum*. Under the inspiration of such ends, out of the effect of the good upon feeling, a joy in taking responsibility can arise; without it, that is, without obligating values, shying away from responsibilities is perhaps to be regretted (since caution can be a bad bargain by purely hedonistic standards), but not to be condemned.

2. *Substantive Responsibility: The Positive Duty of Power*

There is, however, the vastly different concept of responsibility that concerns not the *ex post facto* account for what has been done, but the forward determination of what is to be done; by its command, therefore, I feel responsible, not in the first place for my conduct and its consequences but for the *matter* that has a claim on my acting. For example, responsibility for the welfare of others does not merely "screen" intended actions with respect to their moral acceptability but obligates to actions not otherwise contemplated at all. Here, the "for" of being responsible is obviously distinct from that in the purely self-related sense. The "what for" lies outside me, but in the effective range of my power, in need of it or threatened by it. It confronts this power of mine with its right-to-be and, through the moral will, enlists it for itself. The matter becomes mine because the power is mine and has a causative relation to just this matter. The dependent in its immanent right becomes commanding, the power in its transitive causality becomes committed, and committed in the double sense of being objectively responsible for what is thus entrusted to it, *and* affectively engaged through the feeling that sides with it, namely, "feeling responsible." In this feeling the abstractly binding finds its concrete tie to the subjective will. This siding of sentiment with the object originates not from the idea of responsibility in general but from the perceived right-plus-need of the object, as it affects the sensibility and puts the selfishness of power to shame. First comes the "ought-to-be" of the object, second the ought-to-do of the subject who, in virtue of his power, is called to its care. The demand of the object in the un-assuredness of its existence, on the one hand, and the conscience of power in the guilt of its causality, on the other hand, conjoin in the affirmative feeling of responsibility on the part of a self that anyway and always must actively encroach on the being of things. If love is also present, then responsibility is inspirited beyond duty by the devotion of the person who learns to tremble for the fate of that which is both worthy of being loved and beloved.

This kind of "responsibility" and "feeling responsible" we have in mind, not the empty, formal one of every agent being responsible for his acts,[8] when speaking of "responsibility for the future" as the mark of an ethics needed today. And we must compare it with the motive principles of earlier moral systems and their theories. We come empirically closer to this substantive, goal-committed concept of responsibility by asking (since according to the two different senses of "responsibility" we can say without contradiction that one is responsible even for his most irresponsible deeds) what is meant by *ir*responsible action. We exclude the formalistic sense of "irresponsible" = lacking the capacity for responsibility = not being accountable to begin with.

3. *What "to Act Irresponsibly" Means*

The gambler who puts his whole fortune at stake acts recklessly; when it is not his but another's, then criminally; but when a family depends on him, then irresponsibly, even with ownership undisputed, and no matter whether he loses or wins. The example tells: only one who *has* responsibilities can act *ir*responsibly. The responsibility slighted in this example is of a most comprehensive and enduring sort. The reckless driver is careless for himself, but irresponsible when he thereby also endangers passengers; by taking them on he has assumed, temporarily and confined to *one* trust, a responsibility which he otherwise does not bear for these persons and their well-being. Thoughtlessness, elsewhere innocent and sometimes likeable, is here an offense in itself, even if everything goes well. In both examples there exists a definable, *non*reciprocal *relation* of responsibility. The well-being, the interest, the fate of others has, by circumstance or agreement, come under my care, which means that my control *over* it involves at the same time my obligation *for* it. The exercise of the power with disregard of the obligation is, then, "irresponsible," that is, a breach of the trust-relation of responsibility. A distinct disparity of power or competence belongs to this relationship. The captain is master of the ship and its passengers, and bears responsibility for them. The millionaire among the passengers who happens to be the principal stockholder and director of the shipping company, and can appoint or discharge the captain at will, has on the whole the greater power, but not within this situation. The captain would act irresponsibly if he obeyed the mighty one against his better judgment, for instance, to break a speed record; although in the other relationship (that of the employee) he is "responsible" precisely to him and can be rewarded by him for his obedient irresponsibility or punished for his disobedient responsibility. In the given situation, the captain is the superior and can therefore bear responsibility.

4. *Responsibility, a Nonreciprocal Relation*

It is not entirely clear whether responsibility in the strict sense can exist between those fully equal (i.e., equal in the situation concerned). Cain's counterquestion to God's query concerning Abel, "Am I my brother's keeper?" rejects—not quite wrongly—the (pretended) imputation of responsibility for the independent equal. In truth, God wants to accuse him not of irresponsibility but of fratricide. There certainly are what can be described as relations of mutual responsibility, such as in a dangerous team enterprise (like mountain-climbing), where everyone must rely for his safety upon the others and thus all become, mutually, "their brother's keepers." But such phenomena of solidarity, the standing by one another in a common cause and danger (like comradeship in war, of which soldiers can tell so impressively), belong rather to another chapter of ethics and sentiment; and the true object of responsibility is here after all the success of the common undertaking, not the joys and sorrows of the comrades, among whom I have no advantage singling me out for special responsibility for the others.[9] The brotherhood of purpose is responsible to the purpose; among natural brothers responsibility arises only when one of them falls on evil days or otherwise needs special help—hence again in conjunction with the one-sidedness characteristic of the nonreciprocal relation of responsibility. Such "horizontal" familial responsibility will always be weaker, less unconditional, than the "vertical" responsibility of parents for children, which in regard to them is not specified but global (i.e., extending to everything in them that needs caring for), and not occasional, but continual so long as they are children. Equally continual, therefore, is the peril from neglect of responsibility—a form of "irresponsibility" that involves no positive act of abrogation like that of the gambler and no unethical behavior in the usual sense. This imperceptible, inattentive, unintentional form of irresponsibility, which for this reason is all the more dangerous and not identifiable by any definite act (since it precisely consists in an inactive leaving things alone) will occupy us later in a wider context.

5. *Natural and Contractual Responsibility*

In the parental example just used, we have a case of responsibility instituted by nature, which is independent of prior assent or choice, irrevocable, and not given to alteration of its terms by the participants; and, in that prime example, it encompasses its object totally. Not so a responsibility instituted "artificially" by bestowal and acceptance of a task, for example, appointment to an office (but also that coming about by tacit agreement or the mere fact of competence): this is circumscribed in content and time by the particular task; its acceptance has in it an element of choice, from which one may later resign or be released. Also, in its inception at least, if not in its course, there is some degree of mutuality involved.

More important is the difference that here the responsibility draws its binding force from the agreement whose creature it is, and not from the intrinsic validity of the cause. The tax official's responsibility for collecting taxes is not predicated on the merits of this or any taxation system but on his undertaking the office. Regarding such responsibilities that are merely stipulated and not dictated by an intrinsic claim of the cause, there can well be behavior contrary to duty or negligent of it, but not "irresponsible" behavior, strictly speaking. This term in its strong sense we rather reserve for the betrayal of responsibilities of the independently valid type by which a true good is endangered. Yet even in the case of the tax official, which directly falls in the weak category, we can defend our general thesis that the "ought-to-be" of the cause is primary in responsibility, namely, insofar as the ultimate object of his (as any official's) responsibility, beyond the direct one, thus the true "cause," is the upholding of the loyalty relations in general upon which society and the living-together of men rest: and this *is* a substantive, inherently obligating good. For this good, always unassured of existence and totally dependent on us, the responsibility is as unconditional and irrevocable as any posited by nature can be—if it is not itself such a one. Thus the unfaithful official who directly is only in violation of duty is indirectly, after all, irresponsible as well. Indeed, it may be that some ulterior "natural" responsibility is involved in every artifically instituted one.

Still, the distinction stands and may be restated in conclusion. It is the distinction between natural responsibility, where the immanent "ought-to-be" of the object claims its agent *a priori* and quite unilaterally, and contracted or appointed responsibility, which is conditional *a posteriori* upon the fact and the terms of the relationship actually entered into. In the latter case, even the requisite power, without which there can be no responsibility, is typically generated by the contract itself together with the duty; in the natural case, it is there to begin with and underlies the object's sovereign claim on it. Evidently, in moral (as distinct from legal) status, the natural is the stronger, if less defined, sort of responsibility, and what is more, it is the original from which any other responsibility ultimately derives its more or less contingent validity. This is to say, if there were no responsibility "by nature" there could be none "by contract."

6. *The Self-Chosen Responsibility of the Politician*

There remains, however a third possibility which, in a manner eminently distinctive of human freedom, transcends the difference of natural and contractual responsibility. So far we have found that a good of the first, self-validating order, if and insofar as it lies in the effective range of our power, and all the more, if in any case already impinged on by our activity, engages our responsibility without our choosing and knows no discharge from it. The (at least partly) chosen responsibility of the appointed task has, *per se,* no such commanding good for its immediate object and permits abdication in appropriate ways. But it also occurs that a good of the first order and dignity, that is, one endowed with a "natural" claim, but which of itself lies quite *beyond* one's present range of power, which he thus *cannot* as yet be responsible for at all and may well leave alone—that such a good is *made* the object of a freely chosen responsibility, so that the choice comes first, from the gratuitous and, as it were, presumptuous wish for just that responsibility, and only then procures for itself the power necessary for implementing it; and with it then, indeed, also the duty.

This is clearly an *opus supererogationis,* but familiar enough. What is outwardly visible is the reach for power. The paradigm case is the politician, who seeks power in order to gain responsibility, and supreme power for the sake of supreme responsibility. Power, to be sure, has its own lures and rewards—prestige, glamour, the enjoyment of authority, of commanding and initiating, the inscribing of one's trace in the world, even the enjoyment of the mere consciousness of it (not to speak of the vulgar gains). The motives of the ambitious in striving for it are probably always mixed, and some vanity at least will have played its part already in the self-confidence of the initial choice. But leaving aside the most blatantly selfish tyranny, for which the "political" is merely a pretext, it will be the rule that the responsibility going with the power and made *possible* by it is cointended in the striving for it, and by the genuine *homo politicus* intended in the first place. The real statesman will see his fame (which he may have quite at heart) precisely in that it can be said of him that he has acted for the good of those over whom he had power, that is, *for* whom he had it. This—that "over" becomes "for"—sums up the essence of responsibility.

Here we have a unique privilege of human spontaneity. Unasked, needlessly, without a prior mandate or agreement (which may later add their legitimation), does the aspirant vie for power so as to burden himself with responsibility. The object of the responsibility is the *res publica,* the common cause, which in a republic is latently everybody's cause, but actually only in the limits of the general civic duties. These do not comprise the assumption of leadership in public affairs; nobody is formally bound to compete for public office,[10] usually not even to accept an unsought call to it. But he who feels the calling for it in himself seeks the call and demands it as his right. Public peril in particular, meeting with the conviction to *know* the way to salvation and to be fit to *lead* it, becomes a powerful incentive for the courageous to offer himself and force his way to responsibility. Thus came Churchill's hour in May 1940, when in a nearly hopeless situation he assumed the direction of affairs that no faint-hearted one could covet. Having made the first necessary arrangements, so he tells us, he went to bed in the serene confidence that the right task had found the right man, and slept a restful sleep.

And yet he could have been wrong. If not his estimate of the situation, that of himself might have been in error. If this had later turned out to be the case, history would pronounce him guilty together with his erroneous conviction. But as little as this conviction, in all its sincerity, could serve him as an excuse, as little can the wagering on its truth in the reach for power, which might eliminate better claimants to the task, be made a straightforward moral duty. For no general rule of ethics can make it a duty, on the mere criterion of subjective certainty, to risk committing possibly fatal mistakes at others' expense. Rather must he who wagers on his own certainty take the never excludable possibility of being in error upon his own conscience. For this, there exists no general law, only the free deed, which in the unassuredness of its eventual justification (even in the mere presumption of its self-confidence, which surely cannot be part of any moral prescription) is entirely its own venture. After this moment of supreme arbitrariness, the law takes over again. Having appropriated the ownerless responsibility, its holder is henceforth owned by it and no longer by himself. The highest and most presumptuous freedom of the self leads into the most imperious and unrelenting bondage.

7. *Political and Parental Responsibility: Contrasts*

Now, it is of the utmost theoretical interest to see how *this* responsibility from freest choice and the one most under the dictate of nature, that of the statesman and that of the parent, have, nonetheless, across the whole spectrum at whose opposite ends they lie, most in common and *together* can teach most about the nature of responsibility. The differences are many and obvious. The one responsibility is (as a rule and for a time) to be everyone's, the other for the rare prominent individual only. Object of the one are the few, intimately related fruits of one's own procreation, each counting in its singular identity, yet still unfinished; object of the other are the many, nameless, in themselves already finished members of the preexisting society, who yet are ignored precisely in their individual identities ("without respect of person"). The origin of the one is the direct causation—whether willed or not—by the past act of procreation, conjoined to the total dependence of the fruit thereof; the origin of the other is the spontaneous taking over of the collective interest, as the premise for possible causative acts in the future, conjoined to the—more or less

willing—making over of this public interest to the *negotiorum gestor* on
the part of the many sharing it. Hence, most basic naturalness of the one
responsibility contrasts with utter artificiality of the other; accordingly,
the first is executed in direct and intimate intercourse, the second at a
distance and through the medium of organizational instrumentalities: in
the one case, the object is present to the responsible subject in the flesh,
in the other, only in the idea. Indeed, if the statesman comprises also the
lawgiver, then the most abstract form of responsibility, the most distant
from its real object, stands over against the most concrete form, which
is nearest to its object. In the face of such extreme differences, what
common traits nonetheless make them blend into the one, most integral
and paradigmatic representation of the primordial phenomenon of
responsibility?

III. Theory of Responsibility: Parent and Statesman
as Eminent Paradigms

1. *The Primary Objects of Responsibility Are Other Human Subjects*
    What is common to them can be summed up in the three concepts of
"totality," "continuity," and "future," referring to the existence and welfare of human beings. Let us first take a look at this fundamental relatum,
"human existence." It has the precarious, vulnerable, and revocable character, the peculiar mode of transience, of all *life*, which makes it alone a
proper object of "caring"; and, moreover, it shares with the agent subject
the *humanum*, which has the first, if not the sole claim on him. Every
living thing is its own end which needs no further justification. In this,
man has nothing over other living beings—except that *he* alone *can* have
responsibility *also* for them, that is, for guarding their self-purpose. But
the ends of his fellow sharers in the human condition, whether he shares
these ends himself or merely recognizes them in others, and the ulterior
self-purpose ("end-in-itself") of their being as such, can in a unique manner be included within his own end: responsibility is first and foremost
of men for men, and this is the archetype of all responsibility. This subject-
object kinship in the relation of responsibility implies that the relation,
though unilateral in itself and in every single case, is yet, on principle,
reversible and includes possible reciprocity. Generically, indeed, the reciprocity is always there, insofar as I, who am responsible for someone,
am always, by living among men, also someone's responsibility. This
follows from the nonautarky of man; and, in any case, at least the primal
responsibility of parental care *everybody* has first experienced on himself.
In this archetypal paradigm, the reference of responsibility to the animate
and to the kindred is most convincingly displayed. Thus, to repeat, only
what is alive, in its constitutive indigence and fragility, *can* be an object
of responsibility. But this is only the necessary, not the sufficient condition

for it. Man's distinction that he alone can *have* responsibility means also that he *must* have it for others of his like—that is, for such that are themselves potential bearers of responsibility—and that in one or another respect he, in fact, always has it. Here the mere capacity is the sufficient condition for the actuality. To be *de facto* responsible in some respect for someone at some time (whether acknowledging it or not) belongs as inseparably to the being of man as his *a priori* capacity for it—as inseparably indeed as his being a speaking creature—and is therefore to be included in his definition, if one is interested in this dubious pursuit. In this sense an "ought" is concretely given with the very existence of man; the mere property of being a causative subject involves of itself *objective* obligation in the form of external responsibility. With this, he is not yet moral, but a member of the moral order, that is, one who can be moral or immoral. Moreover, the fact of particular, manifest responsibilities, with their several concrete obligations, is not yet identical with the abstract "ought" which from the ontological claim of the idea of man (see p. 43 f.) secretly goes out to all men and searches among them for its executors or guardians.

2. *Mankind's Existence—the First Commandment*

Of man's prerogative among the claimants on human responsibility there is this to say: it has nothing to do with a balance sheet of mankind's performance on earth, that is, whether it has so far made itself deserving of the preference. The Socratic life or the Beethoven symphony, which one might cite for the justification of the whole, can always be countered with such a catalogue of incessant atrocities that, depending on the appraiser's disposition, the balance can turn out to be very negative indeed. Pity and outrage of the pessimist are not really refutable here; the price of the human enterprise is, in any case, enormous; man's wretchedness has at least the measure of his greatness; and on the whole, I believe, the defender of mankind, in spite of the great atoners like Saint Francis on his side, has the harder case. But such value assessments have no bearing on the ontological issue, as little as the hedonistic balance of happiness and unhappiness (which also tends to turn out negative when—and because—attempted). The dignity of man *per se* can only be spoken of as potential, or it is the speech of unpardonable vanity. Against all this, the *existence* of mankind comes first, whether deserved on its past record *and* its likely continuation or not. It is the ever-transcendent *possibility,* obligatory in itself, which must be kept open by the continued existence. To preserve this possibility is a cosmic responsibility—hence the duty for mankind to exist. Put epigrammatically: the possibility of there being responsibility in the world, which is bound to the existence of men, is of all objects of responsibility the first.

"Existence of a mankind" means simply that there live men on earth; that they live well is the second commandment. The naked ontic fact of their existing at all, in which they had no say, becomes for them the ontological command: that there continue to be such. This, normally unexpressed, "first imperative" is implicit in all the further imperatives (unless these have made nonbeing their aim). Its immediate execution is entrusted to the instinct of procreation, and so it can normally remain hidden behind the particular commands of human virtue, which work out its wider meaning. Only very exceptional circumstances (as today's) may necessitate its becoming explicit as such. But tacitly, it always stands behind the others as their common sanctioning ground. Groundless itself (for there could be no commandment to invent such creatures in the first place), brought about with all the opaque contingency of brute fact, the ontological imperative institutes on its own authority the primordial "cause in the world" to which a mankind once in existence, even if initially by blind chance, is henceforth committed. It is the prior cause of all causes that can ever become the object of collective and even individual human responsibility.

3. The "Responsibility" of the Artist for His Work

Within this generic framework, which primally binds responsibility of any kind, objectwise, to animate beings, inanimate things too can certainly become the "cause" to which responsibility dedicates itself, and this not even (at least subjectively) in the service of a life-promoting end but utterly for its own sake, indeed, with disregard of everything usually going by the name of life-interests. The artist comes to mind. There is such a thing as that hard-to-define, in its own way highest, "responsibility" of the genius for his work, which imperatively takes possession of the one blessed or cursed with the call. Whatever of an "ought" may here be involved turns for him into a "must" that eclipses everything else. Gratification or edification of mortals need not concern him; the egotism of his selfless abandonment to the cause can be complete. This lies beyond morality, and, if the creator destroys his work before anyone sees it, he incurs no guilt toward anyone. Yet even then it could be said that one sense of the ontological experiment which Being undertook with Man is being fulfilled in the fact that such individuals occur, and elucidates itself in what they do, even if it leaves no trace behind. Or perhaps one should say rather that, by such a thing occurring, the open experiment in possibilities receives in retrospect a sense which it could not have had in prospect: for nothing of the kind was written in either stars or genes when the first toolmaker departed from animality, not to mention that no law of survival demanded it. The pure *opus supererogativum* (which morality certainly is not) is in itself a testimony to the transcendence of man, whatever its further effects within the world might be. However, as a part of the world, once it has become this (and for which as a rule it was intended), the work of art exists after all for men only, for their sake, and only as long as such are around. The greatest masterpiece becomes a mute piece of matter in a world without men. And vice versa, without it and the like of it, the world inhabited by men is a less human world and the life of its inhabitants poorer in humanity. Thus the production of the work of art does belong to the world-building deeds of man, and its presence forms part of the stock of the self-created world in which alone human life can be at home. To the artist himself, of course, no such motive as augmenting the artistic stock or furthering the cause of culture is to be imputed; he better thinks of nothing but his work. Regarding then the preservation of his work by others, as a good for mankind (doubtless a duty), *this* does not enjoy the immunity with which its creator, responsible only to his work, could perhaps disregard all other duties. In the famous (in my opinion, perverse) casuistic dilemma of the burning house out of which only one of these two can be saved: Raphael's Sistine Madonna or a child, the morally self-evident decision in favor of the child[11] does not depend upon a comparison of either's "value" for future humanity. (This leaves it morally quite permissible, however, for one to sacrifice *himself* for the work of art, as perhaps the artist has already done in its creation.) To sum up, apart from the scarcely classifiable case of artistic creation, the fact remains that the referent of responsibility—the field of its "causes"— is life, actual or potential and above all human.

4. *Parental and Political Responsibility: Both Are "Total"*

Of such causes we have singled out the two eminent ones of parental and political responsibility, and we named "totality," "continuity," and "future" as the distinctive traits by which they most fully exemplify the nature of responsibility as such. Let us run through them briefly, taking "totality" first. By this we mean that these responsibilities encompass the total being of their objects, that is, all the aspects of them, from naked existence to highest interests. This is obvious for parental responsibility, which really, in time and in essence, is the archetype of all responsibility (and also genetically, I believe, the origin of every disposition for it, certainly its elementary school). The child as a whole and in all its possibilities, not only in its immediate needs, is its object. The bodily aspect comes first, of course, in the beginning perhaps solely; but then more and more is added, all that which falls under "education" in the broadest sense: faculties, conduct, relation, character, knowledge, which have to be stimulated and guided in their development; and together with these also, if possible, happiness. In brief: the pure being as such, and then the best being of the child, is what parental care is about. But isn't this precisely what Aristotle said of the *raison d'être* of the state: that it came into being so that human life would be possible, and continues in being so that the good life is possible? This then is also the object of the true statesman.

The ruler of old took pleasure in having himself regarded as the "father" of his subjects (Papa Czar), implying a kind of tutelage relation which is not inherent in the nature of political life. But a grain of aptness is left to the symbol even in enlightened republics, wherever the chief executive (singular or plural) leads and does not simply carry out the majority will. Leaving aside the dynastic ruler, the "statesman" in the term's full sense has, for the duration of his office or his power, responsibility for the total life of the community, the "public weal." (For "statesman" one may *ad lib* substitute "ruling council.") The way in which he has gotten his mandate is a matter apart. Even usurpation creates responsibility concomitant with the power, and to acquire power for the sake of responsibility can well be the motive for the *coup d'état*. But even if power alone was sought, its possession then objectively carries responsibility with it. Its compass makes it an analogue of parental responsibility. It too extends from physical existence to the highest interests, from security to abundance of life, from good conduct to happiness.

5. *Interpenetration of the Two in Their Object*

No wonder, then, that the two so divergent responsibilities interpenetrate in remarkable ways from the opposite ends of greatest singularity and greatest generality. First, in their object. The education of the child includes socialization, beginning with speech and progressing with the transmission of the entire code of societal convictions and norms, through whose appropriation the individual becomes a member of the wider community. The private opens itself essentially to the public and includes it in its own completeness as belonging to the being of the person. In other words, the "citizen" is an immanent aim of education, thus a part of parental responsibility, and this not only by force of the state's enjoining it. From the other side, just as the parents educate their children "for the state" (if for much more as well), so does the state assume responsibility for the education of the young. The earliest phase is left in most societies to the home, but everything after that comes under the supervision, regulation, and aid of the state[12]—so that one can speak of a public "educational policy." In other words, the state is not content with receiving its citizens ready-made, but chooses to take a hand in their forming. Indeed, it can even, when necessary, assume the child's protection *against* its parents, enforce the discharge of parental duties, and so forth, and do this in precisely those early childhood stages which are normally (except in extreme collectivization: see below) free of public interference. But the prime example, of course, is general compulsory school education, and whatever the official theory might here proclaim (for instance, avoidance of all "bias"), a modicum of ideological indoctrination, to prepare the pupil for his future societal role, is simply inseparable from the transmission of the "instructional matter."

Thus the sphere of education shows most clearly how parental and political responsibilities, the most private and the most public, the most intimate and the most general, overlap (and, for the rest, complement each other) in virtue of the totality of their respective objectives. Admittedly, with extreme collectivization, the totality can go so far on the public side that it fully substitutes itself for the private and abolishes parental responsibility along with parental power. That would be the counterextreme to a primeval condition in which parental or familial power was complete and impervious to outside interference, except for the (then indeed powerful) influence of the general mores. What the abolition of the family as the basic form of generation-spanning human coexistence— the ultimate outcome of total collectivization—would mean in the long run for mankind would remain to be seen. For sure it would mean an enormous shrinking of the private sphere (which, by the way, need not be that of the individual alone) and the near-vanishing of the distinction between it and the public realm. The "statesman" would then have to take care of *everything*. This is *one* meaning of "totalitarianism," which accordingly appears to be inseparable from radical communism. However, this extreme case only magnifies what we assert about the responsibility of the statesman in general and its affinity to that of parents. Indeed, in the light of such trends, we might well say that political history displays a progressive shift of jurisdictional boundaries in favor of the state, that is, the increasing transference of parental responsibility to it. As a consequence, the modern state, be it capitalist or socialist, liberal or authoritarian, egalitarian-democratic or elitist, is becoming in effect ever more "paternalistic."

6. *Analogies of the Two in Sentiment*

a) Not only from the side of the object but also from that of their condition in the subject do the two "total" responsibilities meet. Everybody knows what the subjective conditions are in the parental case: the consciousness of one's total authorship; the immediate appeal of the child's total need for care; and spontaneous love, first as the postpartural, "blindly" compulsive feeling of the mammal mother for the newborn as such, then increasingly, with the emergence of the person, the seeing, personal love for this subject of unique identity. In such choiceless force of immediacy, the subjective condition is as little as the objective replicable in other, less original relations, and the reproductive bond retains a status of primacy that no analogue can equal in self-evidence of responsibility. The statesman is not the originator of the community for which he assumes (perhaps even arrogates) responsibility unto himself; rather its prior existence is the ground for his doing so, and from it he also derives whatever power he may concentrate in his hands. Nor is he, parent-like, the source of sustenance for the collective, but at best the guardian of its continued capacity to sustain itself. More generally, the ruler's responsibility concerns independent beings who in a pinch might manage for themselves. And "love," finally, in the genuine sense cannot be felt for a nonindividual, collective, largely abstract entity. Nonetheless, to take the last and most fundamental point first, there does exist an emotional relation comparable to love on the part of the political individual toward the community whose destiny he wishes to guide to the best, for it is "his" in a much deeper sense than that of a mere community of interests: he is (in the normal case) descended from it and through it has become what he is; he is thus, indeed, not the father, but a "son" of his people and country (also class, etc.) and thereby in a kind of sibling-relation to all the others—present, future, even past—with whom he shares this bond. This fact engenders, as in the family, from which the symbolism is borrowed, something more than merely a recognition of duty, namely, that emotional identification with the whole, the felt "solidarity," which is analogous to the love for the individual. Even solidarity of fate can in terms of sentiment take the place of common descent.[13] If both coincide (a radical contingency the one as much as the other), their combination is overpowering. The fact of feeling then makes the heart receptive to the duty, which of itself does not ask about it, and ensouls the affirmed responsibility with its impulse. It is difficult, though not impossible, to carry responsibility for something one doesn't love, and one rather generates the love for it than do one's duty "free from inclination." Naturally, the partiality of love (which is always particular) can also, and perhaps sometimes must, do injustice to the wider whole of human responsibility that lies beyond its own scope. But assuming responsibility is selective in any case, and the choice for what is close to one's heart is in accordance with human finitude. Thus there is a natural element also within the artificially created *officium* of the statesman, when he—stepping out of the equality of siblings and citizens—assumes for all of them a role which is parent-like in its responsibilities—although natural parenthood has nothing to do with solidarity.

   b) The spectacle of the total *dependency* of the infant, too, has a somewhat more abstract analogue in the political sphere: the general but always perceptually particularized knowledge that the issues of common welfare do not simply look after themselves but require conscious direction and decision, nearly always improvement, and sometimes salvation from disaster. It is, in brief, the insistent knowlege that the *res publica* too exists precariously. Thus, we have here again the fact of vulnerability and peril in that with which sentiment identifies and of which "one" must take care. This "one," meaning abstractly everybody, turns into the self-chosen concrete "I" of the politician who believes that he at this moment knows best what is best for "all," or is best fitted to carry out an existing consensus about it. Whether the belief is right remains forever objectively moot (for his occupying the role prevents the trying out of others), but subjectively this belief belongs inalienably to the total nature of political responsibility as it responds to the call of public necessity. But he who thus responds to public necessity is himself subject to it and rises to the challenge out of the condition of an equal among equals; in the public cause he promotes his very own cause as included in the former. This clearly sets his role off from that of parents who do not share the child's needs but must have outgrown them to be able to minister to them. Beyond this automatically fulfilled condition (as concomitant with procreative maturity), no qualification of special ability is required, whereas the political pretender needs that distinction to legitimate his chosen role.

c) What, therefore, has no equivalent at all in the political sphere is the unilateral and absolute causation of existence, wherein alone, without further supplements, the obligation as well as the qualification for the parental role is grounded, and no analogous feeling unites political responsibility with the parental one. The statesman, however much of a "founder," is himself already a creature of the community whose cause he takes into his hands. He is indebted, therefore, not to what he has made but to what has made him, to the forebears who transmitted the commonwealth down to the present, to the contemporary joint-heirs as the immediate source of his mandate, and to the continuation of the received into an indefinite future. Something of this applies also to the parental role, in qualification of the purely originative relation toward a *de novo* beginning life. But here we are already touching upon the two other characteristics of our two models of responsibility, "continuity" and "future," which almost automatically follow from the characteristic of "totality." The first of these we can treat quite briefly.

7. *Parents and Statesman: Continuity*

Continuity follows from the total nature of responsibility, first in the almost tautological sense that its *exercise* dare not stop. Neither parental nor governmental care can allow itself a vacation or pause, for the life of the object continues without intermission, making its demands anew, time after time. More important still is the continuity of the cared-for *existence* itself as a *concern,* which both responsibilities discussed here must keep in view on every single occasion of their exercise. Nontotal, defined responsibility is limited not only to a single aspect but also to a single time period of such an existence. The ship's captain does not ask his passengers what they did previously or what they will do later—whether they undertake the journey with good or evil intentions, to their salvation or damnation, to the advantage or detriment of third parties. None of this is his concern. As his business is transporting them safely from one place to another, his responsibility begins and ends with their presence on the ship. Or, to take the most frequent example of an at once high and strictly limited responsibility: the responsibility of the physician, begun with the therapeutic relationship, encompasses the curing, relief of suffering, and prolonging the life of the patient. All his other weal and woe lies outside its scope, and the "worth" of the existence benefited or saved is none of its business. And this responsibility, too, ends with the termination of treatment. Total responsibility, however, must continually ask: "What comes after that? Where will it lead?" and at the same time, "What preceded it? How does what is happening now fit into the overall becoming of this existence?" In other words, total responsibility must proceed "historically," embracing its object in its historicity, and this is the true meaning of what we here designate with the element of "continuity." In this regard political responsibility, of course, has an incomparably wider span both backward and forward, corresponding to the longer life of the historical community. We need not elaborate here on the obvious claims of tradition—all that was inherited from the deeds (but also from the omissions and sins) of the forebears—nor on the equally evident claims of a future for the collectivity beyond those now alive. The aspect of the future will occupy us presently by itself. Let us say only that in the continuity through time, an *identity* is to be preserved which is an integral part of collective responsibility.

In parental responsibility, which is focused with such utmost concentration in every instance upon the singular individual, the horizons of responsibility are doubled. The first, and narrower, horizon enfolds the individual becoming of the child, who has his own personal historicity and gains his identity "historically," that is, by way of his own, individual history. Every educator knows of this. But beyond this, and inseparably linked to it, is the communication of the collective tradition, from the first phoneme, and the readying for life in the society. Herewith the horizon of continuity expands into that of the historical world: the one passes into the other, and so educational responsibility cannot help being, even in its most private form, also a "political" responsibility.

8. *Parents and Statesman: The Future*

Above all, however, it is the *future* with which responsibility for a life, be it individual or communal, is concerned beyond its immediate present. In a trivial sense, namely, regarding the progress and outcome of each task at hand, this is true of every, even the most circumscribed, responsibility: the fever curve of the next day, the stretch of the journey yet to traverse, are included within the concern of the moment. But this routine inclusion of tomorrow in today's concern, given with temporality as such, acquires an altogether different dimension and quality in the context of "total responsibility" as we ponder it here. In its indefinite scope, it presents something of a paradox. Children outgrow parental care, communities outlive statesmen. And yet, the pertinent responsibilities, in all of their particular, time-bound tasks, somehow extend into the whole future of their charge, even though that future lies beyond their ken and control. Can the unknowable be included in my duty? Here lies the paradox. For it is the future of the whole existence, beyond the direct efficacy of the responsible agent and thus beyond his concrete calculation, which is the invisible co-object of such a responsibility in each of its single, defined occasions. These occasions, and the interventions they provoke, are each time about the proximate particular, and this lies more or less within the range of informed prescience. The totality that will absorb the long-range effect of the particular decision is beyond such prescience, not only because of the unknown number of unknowns in the equation of objective circumstances, but ultimately because of the *spontaneity* or freedom of the life in question—the greatest of all unknowns, which yet must be included in the total responsibility. Indeed, precisely that in the object for whose eventual self-assertion the original agent can no longer be held responsible himself, namely, the own, autonomous causality of the life under his care, is yet an ultimate object of his commitment. It can be so in one way only: respecting this transcendent horizon, the intent of the responsibility must be not so much to determine as to enable, that is, to prepare and keep the capacity for itself in those to come intact, never foreclosing the future exercise of responsibility by them. The object's self-owned futurity is the truest futural aspect of the responsibility, which thus makes itself the guardian of the very source of that irksome unpredictability in the fruits of its labors. Its highest fulfillment, which it must be able to dare, is its abdication before the right of the never-anticipated which emerges as the outcome of its care. Its highest duty, therefore, is to see that responsibility itself is not stifled, whether from its source within or from constraints without. In the light of such self-transcending width, it becomes apparent that responsibility as such is nothing else but the moral complement to the ontological constitution of our *temporality*.

Where does that leave us practically? For mere mortals, the incalculability of the long-term outcome of any action might seem either to place an impossible strain on responsibility, which could paralyze action, or to provide a facile shelter in the immunity of ignorance, which could excuse recklessness. But the above considerations tell otherwise: in explaining the unknowability and deferring to its cause, they allow us to extract a practical knowledge from ignorance itself. First of all—to say the obvious—as there is no complete knowledge about the future, neither is there complete ignorance, and an agent's *concrete* moral responsibility at the time of action does extend farther than to its proximate effects. How far ahead depends on the nature of the "object" and on our power and prescience. In these respects, there are significant differences between the individual, as the object of child-rearing, and the collective, as the object of statecraft.

IV. Theory of Responsibility: The Horizon of the Future

1. *The Goal of Child-Rearing: Adulthood*

Certain differences in embracing the future between parental and political responsibility leap to the eye. Parenthood deals with persons-to-be in process of becoming that, and this process has its predetermined phases, which must be passed through in their order and time, and its predetermined end, adulthood, when with childhood also parental responsibility ceases. (What takes its place is here not our concern.) All this is known in advance in its structural generality and, in affirmation of the self-active biological dynamics, also willed in advance. Child-rearing thus has a definite substantive goal—the independence of the individual, which essentially includes his capacity for responsibility—and with reaching it (or its imputability) also a definite termination in time. This termination arrives according to its own law and not according to the pleasure of the educator—not even according to the measure of his success, for nature has its hand in the game and grants, for once only, a definite time span within which education must have done its job. Thereafter, the previous object becomes himself a subject of responsibilities, and although precisely this in its undetermined openness was included in the anticipation of parental responsibility, this openness of it just means that the responsibility no longer exists as a task and can only ask itself in retrospect, as the thus emancipated existence goes its own way, whether it has performed its task well or poorly. But whether well or poorly, it had to adapt to the time sequence of organic growth to which also personal growth is tied, so that here historicity and nature deeply interpenetrate in the object.

Thus, to sum up, in child-rearing a definite terminus is set by the object itself. Parental responsibility has maturity for its goal and terminates with it. Adulthood, if its self-assertive powers have not been impaired in advance (a vital "if" indeed), is trusted to make its own new start of coping with the—generally recurrent—conditions of individual life. Each generation stands on itself, repeating the parental precedent in its own way. The aged onlooker may have occasion to doubt if his educative record was blameless, but he has at least nurtured the chance of autonomous life to make up for his sins. "Maturity" just implies this indelible chance, which takes over where parental responsibility ceases. That preordained threshold defines the latter's terminal goal.

2. *Historical Becoming Different from Organic*

No such intrinsic terminus is set to political responsibility or that for mankind as a whole by the nature of its object. The continuous life of communities has no true analogy to the ages of man, and the public collective sphere offers nothing that really parallels the facts of individual ontogeny; in particular, historic becoming defies comparison with organic becoming. Admittedly, this sweeping negation is at odds with influential theories of history and the seductive spell of cherished metaphors. We notoriously love to speak of the youth, the prime, the senescence of a civilization, just as of spring, summer, and autumn in the unfolding of an artistic style. All this is quite fetching and not without its grain of truth, but we do better to resist it. In lieu of a full-dress argument (some of which we reserve for later), let me here just spell out a bit the negative assertion itself. The history of societies, nations, or states—in short, "history"—has no predetermined goal toward which it strives, moves, or should be helped to move: concerning historical entities, we cannot speak of childhood, maturity, old age in any legitimate sense; all organic comparisons, especially those of growth, tempting as they are, do not truly apply and ultimately mislead. Perhaps it helps to state the truism that every society has always been composed of members of every age; control has always rested with adults and most often, especially in early societies, with the oldest, the "elders," as those most mature in experience, knowledge, and judgment. "Primitiveness" of culture (a highly relative term) means neither "young" nor "old"; nor does the "ahistoric," seemingly timeless character of certain societies that stay unchanged. Of an "infancy" of mankind we can speak only mythologically or with poetic license. Paleolithic men, nay, the famous and forever unknown "primal horde," had to meet the harsh demands of their existence, and if they had done so like children we should not be here today. The myths at all times were anything but childish, the rituals anything but playful, magic anything but naive, the fear of the unknown anything but immature, the taboo-ordering of societal relations (kinship systems, exogamy, etc.) anything but uncomplicated or artless; and—not to forget—the technology is *always,* at every stage of evolution, ingenious, and the cunning in the outwitting of nature is probably most often superior to what the modern city-dweller could still muster. The condescension of earlier missionaries and explorers (also slaveholders), who talked of their primitives or "savages" as of children, has long since disappeared from anthropology. But no less should one distrust those who pretend to know about a future destination of their own or every society, about a goal of history, for which all of the past was but a preparation and the present is only a transitional stage. Here it becomes clear whence our position has to expect contradiction: from the political eschatologies of history[14] and from the non-political belief in endless progress. We shall deal with some of their claims later. Our summary thesis is clear: the future (not to speak of its essential unknowability) is no less, but also no more, "itself" and for its own sake than was any portion of the past.[15] Becoming in history, which of course occurs, though by no means incessantly—the becoming of mankind, if you wish—has an entirely different sense than the becoming of an individual from germ to adulthood. Mankind, since it exists (what preceded it in evolution is inaccessible to any apprehension from within), is always "already there" and never still to be brought about.[16] And although *in* its being it is, by force of circumstances and of its own doings, subject to manifold becoming—that is, to always fully human history—mankind is in none of this the object of a programmed schedule of completion, from the unfinished to the finished, from the provisional to the definitive, as its always newborn individuals are. Of mankind it can never be said (except in idle speculation) what it is "not yet," only in retrospect what it *was* not yet at this or that time in the past: for instance, medieval man was not "yet" scientific man, the painting on gold-ground was not "yet" representation of perspective in depth, the nomad was not "yet" an agriculturist. But the being human, though different each time, was not therefore more provisional and "unfinished" at any time than it is today.

3. *"Youth" and "Old Age" as Historical Metaphors*

If, however, we do wish to speak of youth and old age in a collective sense, we must first remember that mankind has been around a long time, and every society encountered today is already old, especially the static, "ahistoric" societies, which characteristically are the "primitive" ones. Biologically, they are all the same age because of an about equally long ancestry. Still, groups can be "young" as such in virtue of being newly formed—for example, by colonizers of a relatively empty space, as in the case of the white settlers of North America, whose arrival indeed marked the "birth" and in that sense the "youth" of a historical entity. With the earth completely peopled, this kind of happening is now a matter of the past, but during that past, as our example shows, it could happen quite "late," in highly advanced history—that is, irrespective of the age of mankind and with the history behind it anything but young. "Young" in a similar sense are also newly founded states, especially of hitherto stateless or recently liberated peoples, where freshness of impulse, inexperience, and daring necessarily combine into a quasi-youthful condition, with the strengths and the weaknesses thereof—where, for example, the comparison with childhood ailments suggests itself. But the spirit of doing things is there from the start as serious and "grown-up" as can be, and as it nearly always is when the collective existence is manifestly at issue. On the other hand, it can happen at any time and in the most advanced civilization that for a brief spell (and always by the fault of statecraft) a childish generation overruns the public stage and later leaves to all of us the bitter harvest of its folly—as we of this century should know full well. Also in more normal comparisons, we can speak of political maturity and immaturity of entire communities, though never with the irreversible sense this would have in organic contexts. However, not such short fluctuations are mostly meant with the historical metaphors borrowed from the ages of man and from the language of biological growth in general, but much longer units of irreversible historical "biography," and there the language becomes seriously wrong.

It is the innocent prerogative of the historian (and of poetic fancy), looking back, for example, with the hindsight of the Empire at the beginnings, to speak of the "infancy" of Rome and to detect in the early seven-hills city the makings, the "germ," of the future greatness. But the politician who, at the time of the Tarquins, or even as late as that of the Licinian agrarian laws, would have proclaimed the dominion over Italy, the Mediterranean, and finally over the *orbis terrarum* to be the destiny of his native city and attempted to direct Roman politics according to this vision, would rightly have been excluded from all public office because of insanity. Still, his prophecy, for all its craziness by standards of verisimilitude and practicality, would at least have been intelligible to his hearers, as it stayed well within the conceptual framework of the times (dominion of one nation over others not being unheard of), so that not a statesman indeed but a seer *could* have uttered the prediction "with frenzied mouth." But at Calvin's time, not even a prophet could have as much as conceived the thought of nineteenth-century industrial Europe, and, if he had done so, nobody would have understood what he was talking about. And yet (who will now deny it?), there *is* a "connection," only not one of foresight or any sort of predictability. Such remote consequences, therefore, cannot fall within the "future-horizon" of political responsibility, for which prescience and causal control—at least in the belief of the subject—are essential preconditions.

4. *The Historical Opportunity: Recognition without Prescience (Philip of Macedon)*

A very different matter is the seizing of historical opportunities, in which there can be full awareness that collective fate is being decided for generations, perhaps centuries. Thus Philip of Macedon clearly grasped, from the condition of the Persian empire, of the Greek world and of Macedonian power, what was now possible; and the success of Alexander's expedition, carefully prepared by him both politically and militarily, proved him correct. Even its (also possible) failure would not have proven him entirely wrong. What the long-term consequences of success would be, he naturally could not foresee either in particulars or in the world-historical total, and every idea which he may have had in this regard was certainly false. But it had indeed become visible that a great turnabout in power relations to the benefit of his monarchy and of a Greece united under its hegemony was beckoning as an attainable prize. Also about the desirability of this prize—concerning which already contemporaries on his own side had different thoughts (e.g., Demosthenes), not to speak of the Persian court and the Asian peoples—there was no doubt in Philip's mind. The retrospective witness, who is plainly unable to marshall the historical alternatives for a balance of desirables, will have to say that an opportunity of the greatest magnitude was seized here with discerning eye and determined action, without forgetting the self-evident truth that chance and good fortune also had to do their share. (E.g., the genius of Alexander was simply not foreseeable in the calculations of Philip, whose assassination could otherwise easily have spelled the end of the dream.)

5. The Role of Theory in Foresight: The Example of Lenin

In this recognition of the moment just illustrated, *theory* had no share. For antiquity, which did not theorize about the political-societal future, this goes without saying. In the modern instance—for example, Lenin's in *his* moment, that of the year 1917—things are apparently different. But only apparently. Marxist theory certainly assured him of the goal, but not of the moment of action for its realization. On the contrary: in this respect, theory had foreseen something quite different; and the doctrine had *post hoc* to be adjusted to the possibility—demonstrated by the deed—of a realization by this path, at this moment, and in this place. Lenin's political genius showed itself precisely in the fact that in the given moment he discovered an unorthodox opening to the orthodox goal (namely, from the most backward end of the industrial-capitalist scale) and against the letter of theory seized the unique chance for the start of the Communist revolution. The success of the action proved the correctness of his view for this critical moment, and nothing but the success could for later judgment distinguish his deed from an adventure. It must remain moot how far beyond this his foresight (true or false) went concerning the whole (particulars being left in any case to improvisation). Certain it is that, for example, the nonarrival of the German revolution forced a comprehensive revision of the accepted program concerning the path, though not of the guiding vision of the end, which by mere reason of its distance (almost like Kant's "regulative idea") is safe from such contamination by reality. But when the road is that long, the first way station within reach becomes itself the goal, gilded by the promise which points beyond it. It is also undecidable whether what was actually achieved in the end was what Lenin really had wanted or would today still regard as that—let alone the desirability of one or the other outcome in itself, which will be forever debatable. But Lenin was surely not mistaken in believing that his deed ushered in a world-historical turn, which gave for generations, if not forever, a new direction to the course of things, and this toward a defined and desired goal. Thus we would have here a case, perhaps the first in history, where the acting statesman could have his eye on vistas of a distant future (at least in the abstract) and thereby also bear responsibility for them, which were completely closed to former statecraft.

6. *Prediction from Analytical-Causal Knowledge*

The role of theory in this example is conspicuous, but intricate. First, obviously and in general, the modern analysis of social and economic causalities is incomparably superior to all earlier knowledge thereof and permits extrapolations into the future which, uncertain as they are, set thinking about the future free from the mere analogies of the past and lead it beyond the iterative induction of experience (i.e., from precedent) into the deduction of the never-been: thus from guessing to calculating the future. At the same time—not unconnected with the better knowledge but as a fact *sui generis*—the power of public *control* over social events, that is, the interventionist causal capacity of the political will (or simply: the power of the state over society), has grown enormously, which in circular turn strengthens the ability to predict and to plan far ahead. Engineer-like designing of future conditions seems to have become possible in principle, and thought models for this are at hand. The obverse of the coin, to be sure, is an ever more intricate and less dissectible complexity of the societal process (thus also of the required models) that must be theoretically and practically mastered. The number of unknowns grows apace with the inventory of the known quantities—a peculiar race between our knowledge and the object's own movement; and as if this were not enough, the psychological feedback of each newly acquired bit of (supposed or real) knowledge adds itself to the unknowns in its own calculation. Whether in the net result the so much better informed predictions have actually become more certain, is moot.[17] But that does not change the fact that today a much larger warp of theoretical knowledge, encompassing a much wider time horizon, is woven into the fabric of collective policymaking (as it ought to be by the dictates of responsibility) than any earlier statesmanship ever dreamed of.

7. *Prediction from Speculative Theory: Marxism*

Added to this, however, and illustrated by our example of Lenin, we have another predictive phenomenon of peculiarly modern vintage, which has little to do with causal analysis in detail: the comprehensive speculative theory of history, which claims to know of something like a pervasive regularity in its object throughout time and from this deduces the predestined future in its main outline. That kind of thing has only become possible with completed secularization, that is, with the principle of total immanence taking the place of transcendent salvation history, and is in historical terms a hardly less novel phenomenon[18] than the soberer causal analysis of concrete societies modeled after the natural sciences. Its outstanding example, of course, is Marxism. Here we have world-historical prognosis on a rational basis—and at the same time, through the unique equation of what must be with what ought to be, a goal-setting for the political *will*, which is thereby itself made a factor in proving the theory true after the latter's preaffirmed truth had first motivated the will on its part. For the political action thus determined, which makes happen what must happen, this closed circuit creates a most peculiar mixture of colossal responsibility for the future with deterministic release from responsibility. We shall deal with the ethics of the eschatology of history in a later chapter. Our present concern is with the role of the theory in projecting the future, which leads to the indicated, immense expansion of the horizon of possible responsibility.

As a theory of the whole of history up to now and yet to come, Marxist theory defines the future in unison with the explanation of the past from one pervasive principle—that is, the future is that part of the total program which after the stages traversed remains still to be brought about. All previous history, which in its essential dynamics is a history of class struggles, will be finally resolved in the classless society, which now stands on the agenda of the total process. And it ought to be thus resolved: the political *will* ought to identify itself with the historical *necessity*, at least among the class charged by the logic of the process with implementing it—the proletariat. The coinciding in this case of self-interest with the universal goal brings it about that for once the interest—itself a part of necessity—takes over the function of the "ought": whereby the irksome chasm between "is" and "ought" is circumvented and the idealism of abstract moral imperative evaded (which by the terms of the theory itself must be impotent). Nevertheless—since said coincidence is not automatically evident, and since, moreover, not everyone acts according to his "best interest" even when evident, and since, lastly, those making it evident to him come here as a rule (if past experience is to be believed) from other classes and thus lack the advantage of the coincidence—that "ought" in the abiding indispensability of its *moral* appeal constitutes a riddle within the determinism of the theory (see below). In any case, given the sweeping character of the theory, the question of truth plays here a more decisive role than in the particularized analytic-empirical knowledge, whose extrapolations—qualified anyway by the coagency of countless other factors—can always be corrected without prejudice to the method as a whole. The integral system, to the contrary, stands or falls with its unqualified correctness, that is, with the truth of its principle and its efficiency both for uniform explanation of the past and prediction of the future. But what does "truth" mean here, and how is it proved?

8. *Self-fulfilling Theory and Spontaneity of Action*

Strictly speaking, the theory can be checked only against the past that preceded its enunciation and thus could not have been influenced by it.

Yet, however well it might fare in this uncontaminated test, the *inference* as such from there to the future is still a leap, which logically can yield no more than a likely hypothesis, but psychologically—considering the resolute commitment it betrays—must as well be inspired by nontheoretical factors from the sphere of sentiment and will: thus, a leap of faith.[19] (Indeed, something of the sort was surely already at work in that particular interpretation of the past, even decisively so, for no unbiased witness will assert that this past *can* be viewed by the intellect *only in this* and no other way.) Thus already at this point of the extrapolation from what has been to what is to come, an element of freedom is involved. But how then is the correctness of the hypothesis or of the faith proved in the further course of things? Not as in the natural sciences through the *fulfillment* of the deduced predictions. For here, where men theorize about men, and publicly at that, *the existence of the theory, being itself a historical factor, changes the conditions of the object of knowledge.* Because it acquires causal power itself in order to help its truth to gain reality, thus with intent contributes to the coming true of its prognoses, it could belong to the self-fulfilling prophecies: its being right in the end would not prove its truth, but rather its power over the minds by which it becomes the cause of particular actions. The theory's turning practical, which in this case is even explicitly provided for in the theory itself, creates therefore verification conditions of a very special sort. (Also falsification conditions: just as success, so can failure be ascribed to the theory's interference in the state of its object—e.g.: had it not been broadcast so widely, capitalism could have defended itself less well against its threat.)

This reasoning, it is true, can be countered with the argument that just that power over the minds already confirms the theory, which can demonstrate the readiness for its acceptance out of its logic of history: even its very appearance at this time, as now historically due, is "predicted" in its own context, so that somehow its mere existence confirms its correctness. We don't deny that a theory which comprises itself, which can explain its own coming-to-be-conceived and even the here-and-now thereof (Hegel's speculative invention)—which, in short, can adduce the fact of itself as proof of its truth—is in a logically impressive position. We forbear to investigate how far the fallacy of the "ontological proof" (for God's existence) might be replicated here,[20] since it does not really matter. For even if we grant the logic of the theory, there remains the fact, alluded to before, that some accepted the truth and some did not, and both either in conformity or in conflict with their interests, so that at least at this point again there intervenes an element of freedom and contingency not accounted for by the theory. But *why* in these cases was the message accepted or rejected? The general tenor of the answer is in no doubt. The theory, apart from all its interpretation of the past, posits a *goal,* whose now arrived possibility, historical necessity, and desirability it has (let us say) demonstrated. Is it too much to assume that the *desirability* in itself, that is, the intrinsic appeal of the goal to personal option, was as a rule the first ground for affirming the theory that posits and legitimates it? Invocation of historical necessity alone will not get anyone to raise a finger. And, needless to say, no court of morals can accept the self-absolution of the political agent who claims that he is but the executor of historical necessity, and that actually not he but rather through him "history" acts. On the contrary, the agent must not only answer for his deed, but also for the conviction which let him see it in this light. This attribution to *him* and his accountability does him greater justice than he allows himself, and defends him against his own self-diminution. For no greater injustice could be done to those who first rallied around the banner of socialism than to deny them the recognition that it was moral outrage, compassion, zeal for justice, and hope for a better, worthier life for the masses which inspired them (and for the most part without anticipating to see the fruition for themselves). The slogan of "scientific socialism," with which the Marxists wished to distinguish themselves from the other, "utopian" socialists, is not to be taken too seriously. "Socialism" remains the key word—an *ideal* which can call forth devotion and then *also* welcome the scientific support. Lenin, Trotski, Rosa Luxemburg are unimaginable without passion in the highest sense of the word—passion for the *good* as they saw it. They were moral natures, committed to a suprapersonal end (with the dangerous conviction, it is true, that the end sanctifies the means), and without this fountainhead of freest spontaneity which defies all calculation of odds, none of the great causes, be its doctrine deterministic or otherwise, would stand much of a chance.

Such passion, cooled by judgment, makes the statesman. Judgment in turn is freedom. Judgment emancipates itself from the prescription of theory. In the case of Lenin, as we saw, judgment selected the unorthodox moment for action. From there on, there was no turning back, and the obstinate course of events dictated to him and to his successors the ever-new exercise of their free, frequently clouded, judgment. Had everything followed the letter of the book, which indeed in this case had traced out the road as well as the goal, there would have been no need for statesmanship: the incumbent in power, a mere functionary, would have merely had to look it up. The history of postrevolutionary Russia, perhaps the greatest instance of professed literalist piety at the top in the history of politics, demonstrates the unofficial contrary all the more impressively. Theory, officially always upheld as sacrosanct, is currently adapted to the recalcitrance of reality (a special sort of hermeneutics in the service of statecraft), and detour becomes the only passable road to the—somehow cleaved to—goal. The detour, however, is the child of circumstance and not of the program. For example, industrialization as a socialist achievement and for the sake of socialism was not provided for in the theory. On the contrary, this had viewed socialism as the dialectical fruit of an industrialism brought to perfection by capitalism.[21]

The question of "correctness" of doctrinal interpretation in this ritual interests no one except the official exegetes. To what extent it would have interested Lenin we don't know. But whether he was more of a pragmatist or more of a dogmatist scarcely makes a difference for the observation that even to his genius, despite his knowing about the enormous *extension* of his undertaking into the future, most of what actually eventuated was impossible to foresee. The one paradoxical certainty here is that of uncertainty. It signifies that the always unexpected, essentially *not*-to-be-anticipated, side of human affairs will never allow the statesman to become dispensable, not even in the "realized" classless society—so that of the once often-cited washerwoman who after work can conduct the business of state we can confidently say: Here Lenin erred. (This, let us note, is itself a prediction!)[22]

V. How Far Does Political Responsibilty Extend into the Future?

1. *All Statesmanship Is Responsible for the Possibility of Future Statesmanship*

How then do things stand regarding the *length* to which political responsibilty can extend into the future? Unlike parental responsibility, it has no point of termination fixed by the nature of the object, and major deeds in its field are apt to create facts never to be undone, constraining the options of all posterity. Abstractly, therefore, responsibility is here endless: power and prescience of the agent alone limit its concrete span. If the two were coextensive, there were no ethical dilemma in wielding the one informed by the other. But that preestablished harmony between the power actually wielded and the predictability of its long-term effects does not exist.[23] Political responsibility is plagued by the excess of the causal reach over that of prescience, so that, in the major cases, it always takes more upon itself than the agent can formally still be answerable for. The consequences of the single action enmesh with the immensity of strands in the causal fabric of the whole, which defies analysis even for the now, and exponentially so into the future. Dispatched into that interplay, the original intent may become immoderately magnified or—just as well—obliterated, and almost surely distorted; in any case, it is set adrift. This amounts to no more than saying that the course of history is unpredictable—and with it the afterlife of political decisions of the moment. Any long-range prognosis is at best an informed guess (usually proved wrong by events: almost every purpose is destined to become estranged from itself in the long run). The spectator, according to his temperament, may be thrilled or chastised by the perennial surprises of the historical drama; the actor must still wager on his guess and live with its uncertainty.

Nonetheless—remembering what we said before—even the most skeptical estimate of historical prognosis leaves at least one basic certainty, itself a prognosis: that political spontaneity will remain necessary at all times, precisely because the excessively intricate web of events will, in principle, never conform to plan. From this there follows a highly general, but by no means empty, *imperative* precisely for the statesman whose action consciously has this enormous causal thrust into the distant unknown, namely, to do nothing that will prevent the further appearance of his like; that is, not to plug up the indispensable, though not calculable, wellspring of spontaneity in the body politic from which future statesmen can arise—therefore, neither in the goal *nor on the road to it* to create a condition in which the potential candidates for a repetition of his own role have become lackeys or robots. In brief, *one* responsibility of the art of politics is to see to it that the art of politics remains possible in the future. Nobody can say that this principle, a knowledge wrested from ignorance, is trivial and not capable of intentional violation (which is one of the criteria of the nontriviality of a principle). The general principle here is that any total responsibility, with all its particular tasks and in all its single actions, is always responsible also for preserving, beyond its own termination, the *possibility* of responsible action in the future—that is, for preserving its own preconditions.

Contemporary civilization has lent a new edge to these considerations. If political action has always been beset by the excess of causal reach over that of prediction and so was never free of an element of gambling, today's global technology has raised the stakes immeasurably and, at the same time, has only widened the gap between the power actually wielded and the predictability of its long-range effects. To be sure, the time span of informed planning has lengthened greatly with the aid of science and its analytical tools, but the span of objective responsibility even more so with the runaway momentum of the novel things set afoot with the same aid. Novelty itself, this greatest boon, has the cost of denying to prediction the benefit of past precedent. Thus the discrepancy between the tremendous time-reach of our actions and the much shorter reach of our foresight concerning their outcome is almost bound to grow bigger as we go on with "big technology." Yet no pleading of ignorance will avail the daring innovators. Possibilities discerned by scientifically schooled imagination take the place of familiar experience in distant anticipation, which has become a moral duty. And among them is for the first time, as a realistic danger of progress, the quenching of future spontaneity in a world of behavioral automata. The point is that the changed scale and content of human action have put the whole human enterprise at its mercy.

2. *Near and Distant Horizons under the Rule of Permanent Change*

What else can be said concerning the time span of political responsibility? Naturally it is always concerned in the first place with the proximate, for the need of the moment demands its remedy, as the opportunity of the moment demands to be seized. Yet farsightedness is part of even this direct coping and is still more required, with novel scope, by the magnified causal range of *modern* actions. But the compass of this farsightedness, as the preceding discussion has shown, has two different horizons: the nearer one, within which—with the analytical knowledge at our disposal and the extrapolations it permits—the effects of single undertakings (e.g., raising or lowering taxes) can be calculated more or less hypothetically beyond the immediate situation; and the farther horizon, in which the momentum of what is now begun leads onward into the cumulative magnitudes of reciprocal interaction with *all* the factors of the human condition. Here, with the many unknowns in the tangled total, nothing really conclusive can still be extracted—except for these two evident aspects: that certain *possibilities* (eventualities) which we can causally construct will "then," if they should materialize, be beyond control; and that their enormous *order of magnitude* will affect the entire destiny of man. This "outer horizon" of uniquely modern power and thus responsibility (with its peculiar dangers of even a best-intentioned belittling of man) we have so far merely indicated and here again leave for later, when we shall examine some such—projectively already traceable—possibilities in their ethical implication for today—to help our groping for an ethic that befits the excess of our power over our knowledge: *this* at least is known to us and must be integrated in our political responsibility.

Concerning the nearer horizon we pointed out that even this today exceeds by far what was accessible to earlier statecraft and to human planning in general. But here we must not overlook a paradox. On the one hand, we know more of the future than our premodern ancestors; on the other hand, we know less. More, because our causal-analytic knowledge with its methodical application to the given is much greater; less, because we must deal with what is constitutionally a state of *change,* while those before us were dealing with what was (or appeared to be) an abiding state. They could be sure that custom, sentiment and opinions, relations of rule, forms of economic life, presence of natural resources, technology of war and peace would not be much different in the next generation from what they were for them. We know, if nothing else, that most of these will be changed. It is the difference between a static and a dynamic situation. Dynamism is the signature of modernity. It is not an accident, but an immanent property of the epoch, and until further notice it is our fate.[24] It bespeaks the fact that we must always figure on novelty without ever being able to figure it out; that change is certain, but not what the changed condition will be. Future inventions and discoveries, for example, cannot be anticipated and allowed for. Only the fact that there will continually be some, and among them some of great, occasionally even of revolutionary significance, is close to certain. But upon that one can base no calculations. This unknown x of permanent innovation haunts every equation. All the projections, which we have developed into such an art with the help of our analysis and of the computer, stand under this caveat. They tell us more and tell it more precisely and farther ahead than earlier appraisals of the future were able to do, but they also must leave more of it open. They run for example as follows: It is foreseeable on the basis of current data and trends (the dynamics as such is allowed for) that the energy supply situation will be such and such in 1990, and in the year 2000 approximately such and such: certain possible advances in a technology that is only just developing, for example, nuclear—whose acceleration is then stimulated by the foregoing calculation—can improve the picture to that and that extent. Experience shows that, given sufficient exertion, such advance can in general be counted on: not indeed with complete certainty, but with enough of it at any rate to command the effort and (where the object is of the right importance) to warrant the risk of even expensive failures. With this response prediction turns into application. The purpose of all predictions is in any case that they be translated into practical politics, namely, in the sense that the actions induced by them will promote or prevent their coming true. Prevention ranks foremost of the two, as the prediction in the sense of warning is naturally and rightly a stronger motive for the exertions of statecraft, surely a more compelling command to responsibility, than the call of promise. At any rate, we have reached this point today, of which we have spoken at some length before (see Chap. 2). In that sense we have to understand, for example, the function of the population estimates for the next decades and into the coming millennium. Those results of the forecast which, by virtue of the quantities already in the running, are unalterable (except through mass extinction) call for early preparations to meet the food requirements, etc., of the numbers then on earth, without havoc to the environment. What beyond that fixed part is still open to influence calls for a policy of timely deflection of the curve away from the direction to disaster. The prophecy of doom is made to avert its coming, and it would be the height of injustice later to deride the "alarmists" because "it did not turn out so bad after all." To have been wrong may be their merit.

3. The Expectation of Scientific-Technical Advances

To return once more to the advance budgeting of future progress, this is necessarily a twilight zone, in which the boundaries of the permissible, that is, the responsible, cannot be sharply drawn. Methodical progress in well-established areas of knowledge almost belongs to the routine of the scientific-technical complex and can be quite consciously (e.g., by money grants) steered in desired directions. Beyond this, as the history of research shows, we can expect from time to time so-called breakthroughs, once work toward them is set in motion after theory has pointed in that direction and attested the possibility in principle (as today in regard to controlled nuclear fusion). But one cannot allow for them in one's planning. To be sure, the not groundless hope for such "breakthroughs" and their continued occurrence in general may well be admitted into the philosophical weighing of the chances in the great wager which the human enterprise as a whole has become. But the statesman, who in a specific case may share the hope, must not gamble if he can avoid it, although sometimes he must. He does not have to gamble in this case and yet can include in his promotional care what has no more, and perhaps less, than an even chance. For the concretely hoped for "breakthroughs" are after all something upon whose threshold science hovers, and, as in the case of the more routine advances, all sorts of things can be done to help it along.[25] In this manner, even that upon which one is not allowed to wager, let alone to reckon, does become an object of forward-looking politics. Or, if you will, the "wager" is staked with surplus means only, not with the substance, the common weal itself, which political planning is about: concerning this, the gambling chance of winning that side bet must not be allowed to play a role other than being welcome when it happens, but no injury when it does not. What we mean is well instantiated by the entirely unspecified, nonanticipatory manner in which the state supports "basic research," that is, pure theory, where no goal is defined and only "something" in general may spin off which might at some time be of use to some practical public interest. A more indeterminate and yet realistic horizon of political responsibility can hardly be imagined.

A very different matter are the expectations of miracles instilled by wish and need, often nourished by a superstitious faith in the omnipotence of science. Of this class is the expectation that entirely new kinds of energy sources will be discovered or entirely new reserves of known ones—that in general there is no end to the welcome surprises of progress, and one or another of them will get us out of the pinch just in time. The possibility indeed is not to be excluded after the lesson of the last hundred years, but to build upon it would be utterly irresponsible.[26] It would be no less irresponsible, however, to build on the prediction that man can get accustomed to anything (or be induced to so accustom himself), although *this* prediction is quite probably correct and in fact, if life principally means adaptation, represents the best and horribly reliable survival insurance which the apostles of unstoppable technological transformation of the conditions of life have to offer. We contend that to build upon this certainty (here conceded) is at least as irresponsible as was, in the previous class of examples, to rely on the uncertain. For here the question is not: Will it work? (One must fear that it will.) It is rather: To what is man *permitted* to accustom himself? To what is it permissible to force or allow him to get accustomed? What conditions therefore may one allow to arise as the dictates to which man's adaptation must bow? These questions bring the *idea* of man into play. It too belongs to the responsibility of the statesman; indeed, it is the ultimate and current content thereof at once, the core of its "totality," the true horizon of its "futurity." We shall say more on that later.

4. *The Generally Expanded Time Frame of Today's Collective Responsibility*

From all of this it follows that, while today there is as little a recipe for statecraft as there ever was, the time spans of responsibility as well as of informed planning have widened unprecedentedly. The excess of the first over the second, which is the moral correlate to the excess of causally effective power over prescience, has been emphasized before and will concern us further. But even the range of concrete, well-defined goal-setting, modest in relation to any distant, "utopian" goal, has assumed new dimensions. "Five-year plans," when the political regime provides the manipulative premises for them, are almost part of the daily fare, and as a rule already have from the start their repetition from the next level in view. The leaders of newly freed ("developing") nations can set for them the goal of catching up with the advanced industrial nations and allow for this two or more generations. Although in this case proven models are imitated and the risks of a novel conception avoided, and even the stages of the way are fairly well mapped out in advance, there are unknowns enough in the computation, and the surprising thing would be if the planners were spared surprises (usually of the unwelcome sort). Naturally the needs of the moment will always have priority, except under the most pitiless regimes which for the sake of the final goal are willing to sacrifice whole segments of their own populations. But enough of these familiar facts of common knowledge.

The crucial point in all this is that the nature of human action has changed and with it the focus of ethical theory. For, reflecting on everything—on the magnitude of our novel powers and the novelty of their products, their impact on the human condition everywhere, and the dynamism they let lose into an indefinite future—we must see that *responsibility* with a never known burden and range has moved into the center of political morality. This is why we make the present effort to clarify the phenomenon of responsibility as such. Here we also have an answer to the question, raised early in this chapter (p. 87), why "responsibility," for which we claim this central place, lacks that prominence and has largely been ignored in traditional ethical theory. Both the fact and the explanation are interesting.

VI. Why "Responsibility" Was Not Central in Former Ethical Theory

1. *Narrower Compass of Knowledge and Power: The Goal of Permanence*
The fact is that the *concept* of responsibility nowhere plays a conspic-uous role in the moral systems of the past or in the philosophical theories of ethics. Accordingly, the *feeling* of responsibility appears nowhere as the affective moment in the formation of the moral will: quite different feelings, as I have indicated, like love, reverence, etc., have been assigned this office. What is the explanation? Responsibility, so we learned, is a function of power and knowledge, with the mutual relation of these two not a simple one. Both were formerly so limited that, of the future, most had to be left to fate and the constancy of the natural order, and all attention focused on doing right what had to be done now. But right action is best assured by right being: therefore, ethics concerned itself mainly with "virtue," which just represents the best possible being of man and little looks beyond its performance to the thereafter. To be sure, rulers looked to the permanence of the dynasty, and republics to that of the commonwealth. But what was to be done to this end consisted essentially in strengthening the institutional and social orderings (including their ideo-logical supports) that would assure such permanence, and moreover in the right education of the heir (in a monarchy) or of the coming citizens (in a republic). What is being prepared is always the next generation, and later ones are seen as its repetition—generations that can live in the same house with the same furnishings. The house must be well built to begin with, and to its preservation is also directed the concept of virtue. Wher-ever the classical philosophers, to whom we owe the science of the state, reflected on the relative goodness of constitutions, a decisive criterion was durability, that is, stability, and to this end a right balance of freedom and discipline seemed the proper means. The best constitution is that which is most apt to last; and virtue, in addition to good laws, is the best guarantee of lasting. Therefore, the good constitution must through itself promote the virtue of the citizens. Accordingly, the true good of the individual (not necessarily, to be sure, of *all* individuals) and the pragmatic good of the state coincide, and this coincidence makes the state an in-trinsically moral and not merely utilitarian institution. The virtuous citizen will cultivate his best faculties (for which freedom is required) and be ready to place them at the service of the commonwealth wherever nec-essary, and withal can also enjoy their possession and exercise in them-selves as his own fulfillment. The polity will continuously profit thereby without preempting the place of personal happiness. All virtues—modes

of personal excellence—display this dual aspect. Courage provides the state with defenders against external enemies; ambition with candidates for the highest offices; prudence among the citizens restrains the state from reckless adventures; moderation curbs the greed which can lead to them; wisdom turns the gaze to goods whose possession is not preemptive and thus cannot become objects of strife (this was later thoroughly changed by the "one true religion"!); justice, which "gives to each his own," prevents or lessens the grudges of inequity which can lead to rebellion and civil war. Justice, in particular, is eminent among the conditions of stability and is accordingly emphasized, but equally for its being a form of personal excellence. (Never is the shaking of the whole edifice recommended for the sake of absolute justice: it is a virtue, i.e., a form of conduct, not an ideal of the objective order of things.) The generally held rule is: what is good for man as a personal and public being now will be so in the future; therefore, the best preshaping of the future lies in the goodness of the present state which, by its internal properties, promises to continue itself. Statecraft accordingly cannot adjourn this goodness to the next or any later generation, but must guard it insofar as it is there, and bring it about in its own day insofar as it is lacking. For the rest, one was conscious of the uncertainty of human affairs, of the role of chance and luck, which one could not anticipate, but against which one could arm with a good constitution of the souls and a sound constitution of the body politic.

2. The Absence of Dynamism

The premise for this reckoning with the essentially same that is threatened only by inscrutable fate is, of course, the *absence of that dynamism* which dominates all of modern existence and consciousness. Things human were seen not otherwise in flux than those of nature, that is, like everything in the world of becoming; this flux has no special definite direction, unless toward decay, and against that the existing order must be secured by good laws (just as the cosmos secures its existence by the preservative laws of its cyclical order). For us moderns, therefore, as long as our being stands under the law of perpetually self-generating change, which must always bring forth, as its "natural" product, things and states genuinely new and never existing before, the political wisdom of the ancients does not lend itself to imitation or assimilation. And it is understandable that for those before us, whose present did not throw such a long shadow ahead into the future as does ours, but mainly counted for itself, "responsibility for the coming" was not a natural norm of action: it would have had no object comparable to ours and be considered hubris rather than virtue.

3. The "Vertical," Not "Horizontal," Orientation of Earlier Ethics (Plato)

But the explanation can go deeper than to lack of power (control of fate and nature), limited precognition, and absent dynamism, all of them negative traits. If the human condition, compounded of the nature of man and the nature of the environment, is essentially always the same, and if, on the other hand, the flux of becoming wherein it is immersed is essentialy irrational and not a creative or directional or otherwise transcending process, then the true goal toward which man should live cannot be seen in the "horizontal," in the prolongation of the temporal, but must be seen in the "vertical," that is, in the eternal, which overarches temporality and is equally "there" for every now. This is best exemplified by Plato, still the mightiest countervoice to the ontology and ethic of modernity. He is also the best touchstone for purposes of distinguishing our standpoint, because his "eros," as the emotional incentive toward the good, is of all its competitors the one most determined by the object and least making a virtue of itself—and we have said of the "feeling of responsibility," to which we now give our vote for this crucial office in the agent's subjectivity, that it is always concerned with an object (a "cause") which the agent recognized as a good and a duty—with the amplification however: "a cause in the world," or hyperbolically even: "the cause of the world." And here lies the difference. The object of eros is the good-in-itself, which is not of this world, that is, the world of becoming and time. The eros is striving relatively for the better, absolutely for perfect being. A measure of perfection is perpetuity. Toward this goal, eros already labors blindly in animal procreation, obtaining a token of eternity in the survival of the species. The "ever again," "always the same," is the first approximation to true being. The seeing eros of man surpasses this in more direct approximations: the eros of the wise aims at it most directly. The drive is upward, not forward, toward being, not into becoming. This direction of the ethical quest is based on a definite ontology. So is ours, but the ontology has changed. Ours is not that of eternity but of time. No longer is immutability the measure of perfection; almost the opposite is true. Abandoned to "sovereign becoming" (Nietzsche), condemned to it after abrogating transcendent being, we must seek the essential in transience itself. It is in this context that responsibility can become dominant in morality. The Platonic eros, directed at eternity, at the nontemporal, is not responsible for its object. For this "is" and never "becomes." What time cannot affect and to which nothing can happen is an object not of responsibility but of emulation. The eternal has no need of the former; it waits for beings to participate in it by way of the latter; and its appearing thus reflected in the medium of the world arouses the longing for it. Only for the changeable and perishable can one be responsible, for what is threatened by corruption, for the mortal in its mortality (as to us moderns, characteristically, only a transient and mortal presence can also be an object of *love*). If this alone is left and at the same time our power over it has grown so enormous, then the consequences for morality are immeasurable but still unclear, and this is what occupies us. The Platonic position was clear: he wanted not that the eternal turn temporal, but that by means of the eros the temporal turn eternal ("as far as is possible for it"). This thirst for eternity is ultimately the meaning of eros, much as it is aroused by temporal images. *Our* concern about the preservation of the species, to the contrary, is thirst for temporality in its ever-new, always unprecedented productions, which no knowledge of essence can predict. Such a thirst imposes its own novel duties; the striving for ultimate perfection, for the intrinsically definitive, is not among them.

4. *Kant, Hegel, Marx: Historical Process as Eschatology*

a) The turning-around of the millennial "platonic" position to that dominant today is tellingly exhibited in the quasi-eschatological philosophies of history from the eighteenth century on, precisely because they still retain, in the idea of progress, a residue of the ideal of perfection, of the "highest good." For example, Kant's "regulative idea" guiding toward such final absolutes is insofar an equivalent of Plato's "Idea of the Good" as this too (although a "constitutive" idea) can be practically understood as the limit goal of an infinite approximation. But the axis of approximation has been pivoted from the vertical down to the horizontal; the ordinate has become the abscissa. The goal which is aspired to, for example, the "highest good," lies in the time series that stretches before the subject indefinitely into the future, and is to be progressively approximated through the cumulative, intellectual or moral activity of many subjects along the line. Thus the outer course of history is credited or charged with what in the Platonic scheme was assigned to the inner ascent of the individual; and the share of the individual subject in the overall attainment of the process can, as in all "progress" models, be only fractional as ordained by his accidental place in time. Actually, however, Kant could not yet in full earnest make the historical process the adequate vehicle for the ideal. For *time*, not truly real for him, is of the phenomenal order only, and of the latter's causality it is not to be expected that it will ever bring about, as a general condition, that coincidence of happiness and moral worth which the "highest good" consists in, or that with its indifference to values it will so much as favor any trends in this direction. Hence, here faith had to help out with its moral hope—framed as a "postulate of practical reason"—that the *transcendent* cause (a residual of the vertical order of things!) will with its hidden, nonphenomenal, moral causality outwit as it were the phenomenal-physical one through the latter's own working, so that the moral will in the world be not in vain. The secularization is here still half-hearted, and the subject, in conforming to the regulative ideal, can at least regard his moral conduct *as if,* beyond possessing its inherent quality, it also contributed to the moral advance of the world. His is, if you will, a fictive, noncausal responsibility, which can ignore the probable course of this-worldly events and yet endow the individual deed with a quasi-eschatological horizon.

b) Hegel then took the step to radical "immanentization." The "regulative" idea, working *through* the willing and acting subjects unbeknownst to them, becomes in history "constitutive," and time, not at all a mere appearance, becomes the true medium for the process of its realization, which occurs as the intrinsic movement of the idea. No longer a mere postulate or "as if," it now is the self-assertive, immanent law of events, and Kantian hope turns into Hegelian necessity. The "cunning" of reason acts, not from without, but through the historical dynamics itself and by means of the entirely different intentions of the executive subjects: with the autonomous power of this unerring dynamics, the moral end is in safe hands, and no one is responsible for it, or can become guilty of its possible frustration. Here the principle of self-moving history has found its grand recognition, but that of the subjects' concrete causality is swallowed up in it.

c) Then with Marx came the famous turning "upside down" ("from its head on its feet") of the Hegelian dynamics and at one with it the insertion of conscious action as a coagency in its now due revolutionary leap. The cunning of reason at last coincides with the willing of the agents, who have identified with its hitherto hidden intention, now become manifest. The becoming known of the intention at the right moment to the right subjects was the last act of the cunning, with which it could abdicate as henceforth unnecessary. And although the actors of the revolution, who now consciously assume their mandate, do not determine the direction of the process, whose executors they rather consider themselves, yet they can (and "ought to"!) assist as midwives at its forthcoming birth. Here, for the first time, *responsibility for the historical future in collusion with history's own dynamism* is put with rational intelligibility on the ethical map; and, for this reason alone, Marxism must be again and again our partner in dialogue, as we pursue our theoretical effort at an ethics of historical responsibility. But in believing to know the direction and the goal, Marxism is still heir to the Kantian regulative idea, which only is stripped of its infinitude and wholly transposed into the finite, and also— with the Hegelian "immanentization"—redeemed from its being apart from worldly causality: it has been appointed the logical law of its dynamics. We post-Marxists (a word perhaps still sounding audacious, and certainly mistaken to many) must see things differently. With technology's having seized power—a revolution this, planned by no one, totally anonymous and irresistible—the dynamism has taken on aspects not contained in any

earlier idea of it and not foreseeable by any theory, Marxist or other. It
now has a direction which, instead of to a fulfillment, *could* lead to a
universal disaster, and a tempo whose frightening exponential acceleration
is apt to escape every control. Thus threatened by catastrophe from the
very progress of history itself, we surely can no longer trust in an immanent "reason in history"; and to speak of a self-realizing "meaning"
of the drift of events would be sheer frivolity. On the contrary, we must
take the forward rushing process in hand, in a wholly new way and *without
a known goal*. This relegates all former conceptions to obsolescence and
charges responsibility with tasks by whose measure even the great question which has agitated minds for so long—whether a socialist or individ-
ualist society, an authoritarian or free one, would be better for man—
changes to the more contingent question of which would be better suited
to deal with the coming situtions—a question of expediency, perhaps even
of survival, but no longer of "Weltanschauung."[27] This remark, however,
is not our last encounter with Marxism.

5. *The Current Reversal of the Dictum "Thou Canst Because Thou Ought"*

The ethical novelty of our situation may be further illustrated by a
comparison with Kant's dictum, "Thou canst because thou ought." As
we reiterate over and over again, responsibility is a correlate of power,
so that the scope and kind of power determine the scope and kind of
responsibility. When power and its constant exercise grow to certain
dimensions, then not only the magnitude but also the qualitative nature
of responsibility changes, with the effect that the deeds of power generate
the *contents* of the "ought," which thus is essentially in answer to what
is being done. This reverses the usual relationship between "ought" and
"can." No longer, what man ought to be and to do (the command of the
ideal) and then either can or cannot, has primacy: but rather what he in
fact already does because he has the power and with it the incentive
thereto; and the duty springs from the deed already underway: it is made his
duty by the stretching causal fate of his actions. Even the ideal to be obeyed
emerges in the process. Kant said: You can because you ought. Today
we must say: You ought because you act—which you do because you
can; which means, your exorbitant capacity is already at work. Obviously,
a different sense and object of capacity is meant in either case. In Kant
it is the ability to subordinate inclination to duty, and this noncausal, inner
ability is generally to be presupposed in the individual, to which alone the
duty addresses itself. (In the collective, to be sure, should this become
the addressee of duties, that ability is highly doubtful, and governmental
coercion is needed.) In our counterdictum, the "ability" means that of
releasing causal effects into the world, which then confront the "ought"
of our responsibility. If with these effects the conditions of existence in

general are jeopardized, then it could be that for a while the higher aspirations for perfection, for the best life, indeed even for the "good will" (Kant) must step back in ethics behind the more vulgar duties which our equally vulgar causality in the world lays upon us. Nobody can say at this moment whether, in the future, Plato's way will not once again be eligible, and we must leave it open whether it may not be more adequate to the truth of being than ours. For the time being, the horizontal dynamics we have unleashed ourselves "has" us by the scruffs of our necks. Even the suspicion that what I called the abolition of transcendence may have been the most colossal mistake in history, does not relieve us of the fact that, now and until further notice, responsibility for what has been set afoot and is kept moving by ourselves, takes precedence before everything else.

6. *Man's Power—the Root of the "Ought" in Responsibility*

With what we have learned on our tortuous journey through the landscape of responsibility, we have also gained an answer to the question which we found obstructing our path at the very beginning (p. 82) and called "the critical point of moral theory," namely, *what passage leads from willing to obligation:* from willing, which in every case, just by pursuing whatever end, actualizes nature's purpose of purposiveness in general and thus is a "good" in itself, to the "ought" which commands or forbids it particular purposes? The transition is mediated by the phenomenon of *power* in its uniquely human sense, in which causal force joins with knowledge and freedom. "Power" as purposive causal strength is in evidence throughout much of animal life. Great is the power of tigers and elephants, greater that of termites and locusts, greater still that of bacteria and viruses. But it is blind and unfree, although driven by purpose; and it finds its natural boundary in the counterplay of all the other forces which carry on the natural purpose just as blindly and choicelessly and in the process hold the manifold whole in symbiotic equilibrium. It can be said that here the natural purpose is administered severely but well, that is, the intrinsic task of being fulfills itself automatically. Only in man is power emancipated from the whole through knowledge and arbitrary will and only in man can it become fatal to him and to itself. His capacity is his fate, and it increasingly becomes the general fate. In him, therefore, and in him alone, there arises out of the willing itself the "ought" as the self-control of his consciously exercised power: and first of all with reference to his own being. Since in him the principle of purposiveness has reached its highest and self-jeopardizing peak through the freedom to set himself ends and the power to carry them out, he himself becomes, in the name of that principle, the first object of his obligation, which we expressed in our "first imperative": not to ruin (as he well *can do*) what nature has achieved in him by the way of his using it. Beyond this commitment to himself, he becomes the custodian of every other end-in-itself that ever falls under the rule of his power. We omit here what lies beyond these duties of guarding and preserving: obligations to ends which none other than he first *creates* as it were out of nothing. For creativity lies outside the tasks of responsibility, which extends no further than to making it possible, that is, to keeping intact its ontological premise, the being of man as such. This is its more modest, but more stringent duty. In sum: that which binds will and obligation together in the first place, *power,* is precisely that which today moves responsibility into the center of morality.

VII. Parent-Child Relation: The Archetype of Responsibility

To conclude these very incomplete reflections on a theory of responsibility we return once more to the timeless archetype of all responsibility, the parental for the child. Archetype it is in the genetic and typological respect, but also in the epistemological, because of its immediate evidence. What has it to tell us?

1. *The Elemental "Ought" in the "Is" of the Newborn*
The concept of responsibility implies that of an ought—first of an ought-to-be of something, then of an ought-to-do of someone in response to the first. The intrinsic right of the object is prior to the duty of the subject. Only an immanent claim can objectively ground for someone else an obligation to transitive causality. The objectivity must really stem from the object. Thus all proofs of validity for moral prescriptions are ultimately reduced to obtaining evidence of an "ontological" ought. If the chances for this were not better than those of the famous "ontological proof" for the existence of God, the theory of ethics would be in a bad way, as indeed it is today. For the crux of present theory is just the alleged chasm between "is" and "ought," which can only be bridged by a fiat, either divine or human. Both are highly problematic sources of validity: the divine, because the existence of its source is contested, while its authority is hypothetically granted; the human, because authority is lacking, while existence is the given fact. In either case, moreover, the fiat anyway merely bridges a chasm which ontologically persists, if the theory is right. This reigning theory holds that from no "is" whatsoever, in either its given or possible being, something like an "ought" can ever issue. Premised here is the concept of a naked "is," present, past, or future. Needed, therefore, is an *ontic* paradigm in which the plain factual "is" evidently coincides with an "ought"—which does not, therefore, admit for itself the concept of a "mere is" at all. Is there such a paradigm? Yes, we answer: that which was the beginning of each of us, when we could not know it yet, but ever again offers itself to the eye when we can look and

know. For when asked for a single instance (one is enough to break the ontological dogma) where that coincidence of "is" and "ought" occurs, we can point at the most familiar sight: the newborn, whose mere breathing uncontradictably addresses an ought to the world around, namely, to take care of him. Look and you know. I say "uncontradictably," not "irresistibly": for of course the force of this, as of any, "ought" can be resisted,[28] its call can meet with deafness (though at least in the mother this is considered an aberration), or can be drowned by other calls and pressures, like sacrifice of the firstborn, Spartan child-exposure, bare self-preservation—this fact takes nothing away from the claim being incontestable as such and immediately evident. Nor do I say "an entreaty" to the world ("please take care of me"), for the infant cannot entreat as yet; and anyway, an entreaty, be it ever so moving, does not oblige. Thus no mention also is made of sympathy, pity, or whichever of the emotions may come into play on our part, and not even of love. I mean strictly just this: that here the plain being of a *de facto* existent immanently and evidently contains an ought for others, and would do so even if nature would not succor this ought with powerful instincts or assume its job alone.

"But why 'evident'?" the theoretical rigorist may ask: What is really and objectively "there" is a conglomeration of cells, which are conglomerations of molecules with their physicochemical transactions, which as such *plus the conditions of their continuation* can be known; but that there *ought* to be such a continuation and, therefore, somebody ought to do something for it, that does not belong to the finding and can in no manner be seen in it. Indeed not. But is it the infant who is seen here? He does not enter at all into the mathematical physicist's view, which purposely confines itself to an exceedingly filtered residue of his otherwise screened-off reality.[29] And naturally, even the brightest visibility still requires the use of the visual faculty for which it is meant: it is to this that our "Look and you will see" is addressed. To contend that this beholding of the given in its fullness possesses less truth-value than the evidence of its last residue in the filter of reduction is a superstition which only lives upon the success-prestige which natural science enjoys outside the cognitive domain it has staked out for itself.

It only remains to explicate *what* is seen here: which traits, besides the unquestionable immediacy itself, distinguish this evidence from all other manifestations of an ought in reality and make it not only empirically the first and most intuitive, but also in content the most perfect paradigm, literally the prototype, of an object of responsibility. We shall find that its distinction lies in the unique relation between possession and non-possession of being, displayed by beginning life, which demands from its cause to continue what it has begun. We must show what is singular and paradigmatic in this situation.

2. *Less Urgent Appeals of an "Ought-to-Be"*

a) It is not altogether meaningless, but not very meaningful either, to ask whether the world ought to be, since the answer, be it positive or negative, has no consequences. The world is already there and continues to be; its existence is not endangered, and even if it were, we could do nothing about it. If God created it, then it probably "ought" to be, but we have no part in implementing this ought. In general, concerning that which exists of itself and is in no way dependent on us, a possibly discerned ought-to-be can have significance for our metaphysical consciousness—certainly when, as here, it includes our existence—but not for our responsibility. Something else indeed is the question whether the world ought to be in this rather than that way, for here there might be room for cooperation and hence responsibility on our part; and this refers us to the narrower realm of human causality. However, such a qualitative ought-to-be, if such exists for the world or part thereof (but if for a part, then through it for the whole as well), is anything but immediately evident and must first be brought to light in an ontological argument—of whose innate (and unsurmountable) logical infirmity our own attempt at the beginning of this chapter furnished proof. And in the main, nature anyway takes care of herself and is indifferent to our evaluating approval or disapproval. At least, whatever duties may nonetheless fall on us to come to her aid in this or that respect, they are anonymous and lack the urgency of the moment. "One" should concern oneself with this or that but not especially and uniquely I, and not necessarily today, but maybe starting day after tomorrow or sometime in coming years. For that which abides by its own strength—the world as it is—can outwait the shortlived needs of man and in general leaves itself the chance that "sooner or later" its better state will gain support against the worse. The "better," *nota bene*, is not necessarily something that is yet to come; it can also be what needs preserving from something that threatens to come and would be worse (as, for example, the wantonly caused extinction of higher animal species).

b) What then, over against this self-sustaining and already existent reality, is the status of things that are not yet, that have never been in the world, but could be and only through our agency can come to be? Here, too, the things in question would have to be future *states*, whether of nature or of society, not individual existents (see below). When such a state is seen as valuable and realizable—and it doesn't have to be "the highest good"—then its generation may well become, in proportion to the "ought-to-be" perceived by us, the charge of human responsibility, namely, for the sake of the already existing and known totality which will gain thereby. For itself, nothing prior to actually existing has a right to exist and, hence, a claim on our aiding its coming to exist (unless we hypostatize its possibility to an expectant already-there in a timeless realm). But the "cause" of the world, for whose sake the state ought to be, is (we say it again) opaque to us and must first be proved for the particular case. Above all, the "state" is something general, a *form* of order: the particular and as yet nonexistent individuals that are to compose it can in no way be anticipated, the question whether *they* "ought-to-be" has no meaning before they are, and the existence of none of them—*ad lib* "exchangeable" as far as the formal order is concerned—can be the content of any conceivable responsibility for the future. Indeed, the planning of any system-state (unless immediately realizable), for example, of "society," is only possible on condition that it is independent of the unique identity of its constituents. Thus, while it does make sense to say that there ought to be men in the future, once "man" is there, the question of "who" the future people will be must fortunately remain open, and to say that this one or that one ought to be before he is, makes no sense at all.[30] Similarly one can rightly say that "freedom" (or "responsibility," etc.) should continue to be actual in the world to come, since the ontological possibility thereto has become evident in the fact of it; and with the acknowledgment of this abstract "ought" one can, given the circumstances, also assume a concrete responsibility for it: but what the deeds of this freedom will be in its time—for this, the very nature of what one seeks to ensure utterly precludes to assume responsibility. Or, to take a final example, one can say that there ought to be art and science, after they exist (one could not say so before) and do one's part in making their further existence possible; but the eventual works of future artists, the discoveries of future researchers, are not only indeterminable in advance and thus no possible objects of responsibility—their being beyond planning is even an essential component of what one here feels responsible for (a memento for the endowers of foundations). Now precisely under this proviso of abstractness do also fall the planning and preparing of states of mankind which have never yet been and go beyond all previous conditions—if for the nonce we suppose the right of utopia and the human power equal to it. Thus in all these cases we owe the anonymous future only the general, not the particular—the formal possibility and not the determinate substantive reality. The statesman, to be sure, in the focusing burning-glass of the situation and its urgency, has normally far more concrete issues within such abstract, distant horizons on his hands; yet even he meets with a practical "ought" of that utmost immediacy which brooks no delay only in the rare and most dramatic moments of decision where the being or not being of the community is at issue. However, precisely this is *not* the exception but the incessant rule in the one and primordial counterexample which we now wish to delineate as the *prototype* of every other responsibility.

3. The Archetypal Testimony of the Infant to the
Nature of Responsibility

Against the background of vague responsibilities just sketched, the always acute, unequivocal, and choiceless responsibility which the newborn claims for himself stands out as utterly beyond comparison. The newborn unites in himself the self-accrediting force of being already there and the demanding impotence of being-not-yet; the unconditional end-in-itself of everything alive and the still-have-to-come of the faculties for securing this end. This need-to-become is an in-between, a suspension of helpless being over not-being, which must be bridged by another causality. The radical insufficiency of the begotten as such carries with it the mandate to the begetters to avert its sinking back into nothing and to tend its further becoming. The pledge thereto was implicit in the act of generation. Its observance (even by others) becomes an ineluctable duty toward a being now existing in its own authentic right and in total dependence on such observance. The immanent ought-to-be of the suckling, which his every breath proclaims, turns thus into the transitive ought-to-do of others who alone can help the claim continually to its right and make possible the gradual coming true of the teleological promise which it carries in itself from the first. They must do this continually, so that the breathing might continue and with it also the claim renew itself continually, until the fulfillment of the immanent-teleological promise of eventual self-sufficiency releases them from the duty. Their power over the object of responsibility is here not only that of commission but also that of omission, which alone would be lethal. They are thus responsible totally, and this is more than the common human obligation toward the plight of fellow humans, whose basis is something other than responsibility. In its most original and massive sense, responsibility follows from being the cause of existence; and all those share in it who endorse the fiat of procreation by not revoking it in their own case, namely, by permitting themselves to live—thus, the coexisting family of man. Hence also the state bears a responsibility for the children within its jurisdiction quite distinct from that for the welfare of its citizens in general. Infanticide is a crime like every murder,[31] but a child's dying of hunger, that is, permitting its starving to death, is a sin against the first and most fundamental of all responsibilities which man can incur.

With every newborn child humanity begins anew, and in that sense also the responsibility for the continuation of mankind is involved. But this is much too abstract for the prime phenomenon of utter concreteness we are considering here. Under that abstract responsibility there may have been, let us assume, the duty to produce "a child," but none possibly to produce *this* one, as the "this" was entirely beyond anticipation. But precisely *this* in its wholly contingent uniqueness is that to which responsibility is now committed—the only case where the "cause" one serves has nothing to do with appraisal of worthiness, nothing with comparison, and nothing with a contract. An element of impersonal guilt is inherent in the causing of existence (the most radical of all causalities of a subject) and permeates all personal responsibility toward the unconsulted object.[32] The guilt is shared by all, because the act of the progenitors was generic and not thought up by them (perhaps not even known); and the later accusation by children and children's children for neglected responsibility, the most comprehensive and practically most futile of all accusations, can apply to everyone living today. So also the thanks.

Thus the "ought" manifest in the infant enjoys indubitable evidence, concreteness, and urgency. Utmost facticity of "thisness," utmost right thereto, and utmost fragility of being meet here together. In him it is paradigmatically evident that the locus of responsibility is the being that is immersed in becoming, surrendered to mortality, threatened by corruptibility. Not *sub specie aeternitatis,* rather *sub specie temporis* must responsibility look at things; and it can lose its all in the flash of an instant. In the case of continually critical vulnerability of being, as given in our paradigm, responsibility becomes a continuum of such instants.

How this primal paradigm is not only in terms of self-evidence and content the archetype of all responsibility, but also its initial germ in the generic human condition,[33] whence it spreads and transfigures itself into other, more public horizons of responsibility, will not be traced out here. Something of it, we trust, will be perceptible by implication in the coming discussion of certain of these horizons. Toward these we must now turn.

5
Responsibility Today: Endangered Future and the Idea of Progress

I. The Future of Mankind and the Future of Nature

1. *Solidarity of Interest with the Organic World*

Care for the future of mankind is the overruling duty of collective human action in the age of a technical civilization that has become "almighty," if not in its productive then at least in its destructive potential. This care must obviously include care for the future of all nature on this planet as a necessary condition of man's own. Even if it were less than necessary in this instrumental sense—even if (science-fiction style) a human life worthy of its name were imaginable in a depleted nature mostly replaced by art—it might still hold that the plenitude of life, evolved in aeons of creative toil and now delivered into our hands, has a claim to our care in its own right. A kind of metaphysical responsibility beyond self-interest has devolved on us with the magnitude of our powers relative to this tenuous film of life, that is, since man has become dangerous not only to himself but to the whole biosphere.

There is no need, however, to debate the relative claims of nature and man when it comes to the survival of either, for in this ultimate issue their causes converge from the human angle itself. Since, in fact, the two cannot be separated without making a caricature of the human likeness—since, rather, in the matter of preservation or destruction the interest of man coincides, beyond all material needs, with that of life as his worldly home in the most sublime sense of the word—we can subsume both duties as one under the heading "responsibility toward man" without falling into a narrow anthropocentric view. Such narrowness in the name of man, which is ready to sacrifice the rest of nature to his purported needs, can only result in the dehumanization of man, the atrophy of his essence even in the lucky case of biological survival. It therefore contradicts its professed goal, the very preservation of himself as sanctioned by the dignity of his essence. In the truly human aspect, nature retains her dignity, which confronts the arbitrariness of our might. Ourselves being among her children, we owe allegiance to the kindred total of her creations, of which the allegiance to our own existence is only the highest summit. This summit, rightly understood, comprises the rest under its obligation.

2. *Egoism of Each Species and the Overall Symbiotic Balance*

In the choice between man and nature as the struggle for existence poses it time and again, man indeed comes first, and extrahuman nature, even with its own rights acknowledged, must give way to his superior right. Or, should the idea of anything intrinsically "superior" be questioned here, the simple rule holds that egoism of the species—each species—takes precedence anyway according to the order of life in general, and the particular exercise of man's might *vis-à-vis* the rest of the living world is a natural right based on the faculty alone. In other words, the mere fact of superior power legitimates its use. This has in practice and without reflection been the attitude of all times, when for long the odds were by no means so clear, when often man felt more on the defensive than on the offensive, and when nature as a whole appeared invulnerable, thus in all particulars free for his untrammeled use. Western religion and metaphysics added their sanction of transcendental uniqueness to this anthropocentric bent. But even if the prerogative of man were still insisted upon as absolute, it would now have to include a duty toward nature as both a condition of his own survival and an integral complement of his unstunted being. We have intimated that one may go further and say that the common destiny of man and nature, newly discovered in the common danger, makes us rediscover nature's own dignity and commands us to care for her integrity over and above the utilitarian aspect. A sentimental interpretation of this command is ruled out by the law of life itself, which is obviously part of the "integrity" to be preserved. For encroaching on other life is *eo ipso* given with belonging to the kingdom of life, as each kind lives on others or codetermines their environment, and therefore bare, natural self-preservation of each means perpetual interference with the rest of life's balance. In simple words: to eat and to be eaten is the principle of existence of just that manifold which the command bids us to honor. (Metabolic exchange with inorganic matter alone—the condition with which it all must once have begun and retained by plants—has been forfeited by all animal life.) The sum total of these mutually limiting interferences, always involving destruction in the particulars, is on the whole symbiotic but not static, with those comings, goings, and stayings known to us from the dynamics of prehuman evolution. The hard order of ecology (first seen by Malthus) prevented any excessive predation by a single life form on the rest, any monopoly of a "strongest," and the joint existence of the whole was assured in the change of its parts. Even the increasingly one-sided interference of man was no decisive exception to this rule until quite recently.

3. *Man's Disturbance of the Symbiotic Balance*

Only with the superiority of *mind* and the inordinate powers of technical civilization eventually engendered by it has one form of life, man, lately acquired the ability to endanger all others (and therewith himself too). Nature could not have incurred a greater hazard than to produce man: with his emergence, it potentially upset its internal balance and left it to the gathering momentum of his career to do so actually. That actuality has now come to pass. Any "Aristotelian" idea of a safe teleology of "Nature" (*physis*) as a whole that attends to itself and automatically ensures the harmonizing of the many purposes into one is refuted by this latest turn, whose very possibility even an Aristotle could not yet have suspected. To him, it was *theoretical* reason in man which stood out above nature and which surely did it no harm in contemplating it. As man's faculties were seen to culminate in this contemplative faculty, so his being was seen to culminate in its exercise. And as long as practical intelligence and theoretical intellect went their separate ways, his impact on the balance of things remained tolerable. But the very meaning of "theory" has since changed.[1] Unlike the contemplative intellect of old, its emancipated heir, the aggressive and manipulative intellect bred by modern science and discharged into the administration of things, confronts nature not merely with its thought but with actions of a scope no longer compatible with the unconscious functioning of the whole. In man, nature has disturbed herself and has only in his moral endowment (which we may still ascribe to her) left herself an unsure substitute for the shattered sureness of her self-regulation. It is a terrifying thought that on this fickle ground her cause should now rest—or let us say more modestly: so much of what man can see of her cause. By the time scale of evolution and even the much shorter scale of human history, this is an almost sudden turn in the fate of nature. Its possibility lay hidden in the initial fact of free-roaming knowledge and will that with man had burst into the world, but its reality matured slowly and then suddenly appeared. In this century the long-prepared point has been reached where the danger becomes manifest and critical. Power conjoined with reason carries responsibility with it. This was always self-understood in regard to the intrahuman sphere. What is not yet fully understood is the novel expansion of responsibility to the condition of the biosphere and the future survival of mankind, which follows simply from the extension of power over these things—and from its being eminently a power of destruction. Power and peril reveal a duty which, through the commanding solidarity with the rest of the animate world, extends from our being to that of the whole, regardless of our consent.

4. *The Danger Makes the "No to Not-Being" Our Primary Duty*

Let me repeat: the duty we talk about has become apparent only with the threat to the subject thereof. Previously it would have been senseless to talk about such things. What is in jeopardy raises its voice. That which had always been the most elementary of the givens, taken for granted as the background of all acting and never requiring action itself—that there are men, that there is life, that there is a world for both—this suddenly stands forth, as if lit up by lightning, in its stark peril through human deed. In this very light the new responsibility appears. Born of danger, its first urging is necessarily an ethics of preservation and prevention, not of progress and perfection. In spite of this modesty of aim, its commandments may be rather difficult to obey and perhaps demand more sacrifices than any asked so far for the betterment of the human lot. We said at the beginning of the preceding chapter that man, no longer simply the further executor but also the potential destroyer of the purposive labor of nature, must make its general "Yes" a part of his own willingness to impose the "No" to Not-Being on his power. It is a consequence of freedom's negative power in its present ascendancy that the "permitted and not permitted" has priority over the positive "ought." This is only the beginning of morals and, of course, insufficient for a positive doctrine of duties. Fortunately for our theoretical task, and unfortunately for our situation today, we need not go into the theory of the human good and the "best life," which would have to be derived from a conception of man's "essence." For the moment, all work on the "true" man must stand back behind the bare saving of its precondition, namely, the *existence* of mankind in a sufficient natural environment. In the total danger of the world-historical Now we find ourselves thrown back from the ever-open *question, what* man ought to be (the answer to which is changeable), to the first *commandment* tacitly always underlying it, but never before in need of enunciation: *that* he should be—indeed, *as* a human being. This "as" brings the essence, as much as we know or divine of it, into the imperative of "that" as the ultimate reason for its absoluteness and must prevent its observance from devouring the ontological sanction itself; that is, the policy of survival must beware lest the existence finally saved will have ceased to be human. Considering the severity of the sacrifices that could be necessary, this may well become the most precarious aspect of the ethics of survival that is being imposed on us now: a ridge between two abysses, where means can destroy the end. This ridge must be walked in the uncertain light of our knowledge and in honoring that which man has made of himself in millennia of cultural endeavor. But what now matters most is not to perpetuate or bring about a particular image of

man, but first of all to keep open the horizon of *possibilities* which in the
case of man is given with the existence of the species as such and—as
we must hope from the promise of the *imago Dei*—will always offer a
new chance to the human essence. This means that the "No to Not-
Being"—and first to that of man—is at this moment and for some time
to come the primal mode in which an emergency ethics of the endangered
future must translate into collective action the "Yes to Being" demanded
of man by the totality of things.

II. The Ominous Side of the Baconian Ideal

All this holds on the assumption made here that we live in an apocalyptic
situation, that is, under the threat of a universal catastrophe if we let
things take their present course. On this subject I now have to say a few
things, well-known though they be. The danger derives from the excessive
dimensions of the scientific-technological-industrial civilization. What we
could call the Baconian program—namely, to aim knowledge at power
over nature, and to utilize power over nature for the improvement of the
human lot—lacked in its capitalist execution from the outset the rationality
as well as the justice with which it could have been conjoined. But its
intrinsic and self-reinforcing dynamics, necessarily propelling into extravagance of production and consumption, would probably have over-
whelmed any society, considering the short range of human targets and
the truly unforeseeable magnitude of its success (for *no* society is composed of wise individuals).

1. *The Threat of Catastrophe from Excessive Success*

Thus the danger of disaster attending the Baconian ideal of power over
nature through scientific technology arises not so much from any short-
comings of its performance as from the magnitude of its success. This
success is mainly of two kinds: economic and biological. Their necessarily
crisis-bound mutual relation is by now becoming manifest. The economic
success, long considered alone, meant increased per capita production of
goods, both in quantity and variety, together with the reduction of human
work, thus heightened prosperity for many and even involuntarily heightened consumption for all within the system—ergo, enormously increased
metabolism of the social body as a whole with the natural environment.
This itself had its dangers of overstraining finite resources (disregarding
here those of internal corruption). But these dangers are raised to a higher
power and accelerated by the—at first less visible—biological success:
the numerical swelling of this metabolizing collective body, that is, the
exponential population growth within the geographical reach of the health
benefits of technological civilization, which far exceeds the reach of its
economic benefits and by now covers the globe. This not only adds a new

quantitative dimension to the first process, increasing its tempo and multiplying its effects on the balance of global metabolism; it also very nearly deprives it of whatever freedom it may otherwise have to call a halt to itself. A static population could say at a certain point, "Enough"; but a growing one has to say, "More!" Today it becomes frighteningly clear that the biological success not only may nullify the economical by leading back from the short feast of affluence to the chronic weekday of poverty, but it also threatens mankind and nature with an acute catastrophe of enormous proportions. The population explosion, seen as a problem of planetary metabolism, takes the helm away from the welfare aspiration and may increasingly compel a mankind in process of impoverishment to do just for the sake of immediate survival what for the sake of happiness it was free to do or leave: to ever more recklessly plunder the planet. Inevitably the latter will have the last word when eventually it denies itself to the overdemand. Imagination recoils from the prospect of mass dying and mass killing that will accompany such a situation of *sauve qui peut*. The equilibrium laws of ecology, for so long held off by art, which in their natural state prevent the overgrowth of any one species, will assert their right all the more terribly the more they have been bullied to the extreme of their tolerance. How after this a remnant of mankind will start afresh on a ravaged earth defies all speculation.

2. *The Dialectic of Power over Nature and Compulsion to Exercise It*

This is the apocalyptic perspective calculably built into the structure of the present course of humanity. It must be understood that we are here confronted with a dialectic of power which can only be overcome by a further degree of power itself, not by a quietist renunciation of power. Bacon's formula says that knowledge is power. Now the Baconian program by itself, that is, under its own management, has at the height of its triumph revealed its insufficiency in the lack of control over itself, thus the impotence of its power to save not only man from himself but also nature from man. Both need protection now because of the very magnitude of the power man has reached in the pursuit of technical progress, where the growing power has engendered the concurrently growing necessity for its use—hence the strange impotence to stop its ever-continued and foreseeably self-destructive progress. Bacon did not anticipate this profound paradox of the power derived from knowledge: that it leads indeed to some sort of domination over nature (i.e., her intensified utilization), but at the same time to the most complete subjugation under itself. The power has become self-acting, while its promise has turned into threat, its prospect of salvation into apocalypse. Power over the power is required now before the halt is called by catastrophe itself—the power to overcome that impotence over against the self-feeding compulsion of power to its progressive exercise. After the first-degree power, directed at a seemingly inexhaustible nature, has changed into a second-degree power which wrested control of the first from the user's hands, it is now up to a third-degree power to enforce the self-limitation of the rule that carries along the ruler, before it shatters on the barriers of nature. This third degree means power over the second-degree power, which was no longer man's but power's itself to dictate its use to its supposed owner and to make of him the compulsive executor of his capacity, thus enslaving man instead of liberating him.

3. *The Sought-after "Power over Power"*
From which direction can we expect this third-degree power which reinstates man—and just in time—in the control of his power and breaks its tyrannical automatism? It must, in the nature of the problem, emanate from society, as no private insight, responsibility, or fear can measure up to the task. And as the "free" economy of Western industrial societies is the very source of the dynamics which drifts to the mortal danger, we look quite naturally to the alternative of communism. Can this bring the necessary help? Is it tuned to it? Under this point of view alone, Marxist ethics should be examined today—that is, under that of salvation from disaster, not under that of fulfillment of mankind's dreams. One looks to Marxism because its professed concern is with the future of the whole human enterprise (as it talks of "world revolution"), for the sake of which it dares to ask from the present any sacrifice and, where it rules, can also enforce it. It is at least more difficult to see how the capitalist West could accomplish this. This much is clear: only a maximum of politically imposed social discipline can ensure the subordination of present advantages to the long-term exigencies of the future. However, since Marxism is a form of progressivism and by no means sees itself as an emergency policy but rather as the road to a higher realization of man, the chances it may unintentionally offer toward the now preeminent goal of prevention must be examined in the context of its innate purpose, whose "meliorism" it shares with the general spirit and dominant tendency of the modern age. With the interrogation of Marxism, we take up the question we tabled when it first came up (p. 17), namely, how the ethic of responsibility for the future we have in mind relates to the progressivist ideal.

III. Is Marxism or Capitalism Better Fitted to Counter the Danger?

1. *Marxism as Executor of the Baconian Ideal*
Ernst Bloch's formula "S is not yet P" (where P is the desired and inherently demanded predicate of the subject S, its optimal state, and bringing it about as a universal condition is our task) can serve as a basic

expression of the progressivist world view.[2] The condition in question is that of man. The not-yet-being of P as the state of *man in general* then imports that authentic man is yet to come and that man up to now is not yet him and never was. All previous history is the prehistory of true man, as he can and should be. If we disregard the vague belief in humanity's moral progress through civilization, which defines no program for action (not to speak of the Nietzschean escapade of the somehow promised superman), two practically prescriptive forms of the ideal have been framed in the West: first, the previously discussed Baconian ideal of increasing power over nature; and then, *already building on this,* the Marxist ideal of the classless society. But only the Marxist program, which integrates the naive Baconian idea of dominating nature with that of reshaping society and from the latter expects the definitive man, can be seriously regarded today as the source of an ethic which aims action predominantly at the future and thence imposes norms on the present. One can say that it proposes to bring the fruits of the Baconian revolution under the rule of the best interests of man and thereby to redeem its original promise of an elevated mankind—which promise was with capitalism in bad hands. To this extent Marxism is active eschatology, in which prediction and will have equal share, which envisions a future good of surpassing moral claim, and which stands entirely in the sign of hope. We cannot avoid comparing this mighty conception of a duty to the future—so superior to anything else that today in the public sphere vies for our assent on a purely secular basis—with the entirely noneschatological conception of duties which we believe to emanate from the emerging global plight. Determining their relationship is not a question of abstract correctness but of concrete urgency, although a certain amount of criticism must be devoted to the Promethean arrogance of the utopian ideal in itself.

2. *Marxism and Industrialization*

What is common to both standpoints—the guarding and the promising—besides the "horizontal" orientation as such, is the premise of technical-industrial civilization and its importance as the starting point for any prognosis. With regard to the "promising" standpoint, this must be briefly explained.

It is not by chance that socialism appeared with the beginnings of the machine age and that its scientific accreditation by Marx was based on the condition of capitalism created thereby. Expressed with rough simplification: only this condition made socialization seem *worthwhile,* quite apart from the fact that it also made it appear *necessary* and politically *attainable* in the light of the crisis theory of capitalism and the pauperization theory of the proletariat. The first assertion is safer than the other two, and will hardly be contested. Only modern technology makes possible such an increase of the social product that its just (equal) distribution does not result in general poverty—whereby the feeling of injustice would be remedied but little else. In an economy of want, the just distribution of the insufficient makes only a slight difference in favor of the many, and it even has to be said that under such circumstances the injustice of the concentration of riches and liberties in the hands of few can at least benefit culture, for which in times of primitive technology a terrible price had always to be paid. (What would classical civilization, the fruits of which we would hardly like to miss, have been without the slave economy?) Equal poverty for all, guaranteed by the state, may be morally less revolting than the riches of a few *vis-à-vis* the poverty of the many, but this gain alone would not have lent wings to the socialist ideal and allowed it to make history. Stated brutally: only the magnitude of the prize awaiting the proletariat made the revolution worthwhile. This is entirely legitimate. It seems somewhat at odds with this reasoning that just where the prize already in existence was the highest—that is, in the advanced industrial countries—the masses so far have not chosen this way, and that today, contrariwise, only in the poorest countries does socialism recommend itself as a means of creating that very prize after the capitalist model. But the prize is at least in sight through that model, and the anomaly of Marxism winning in backward rather than advanced societies does not alter the fact that the proof of material surplus already furnished by modern technology, if not at home then elsewhere, is an important factor in the modern socialist ideal.

Indeed, wherever socialism came to power, industrialization was the hallmark of its actual and resolute politics. It is therefore still true today that Marxism, progressivist as it was from the beginning, born in the sign of the "Principle of Hope" and not the "Principle of Fear," is no less dedicated to the Baconian idea than its capitalist rival, with which it competes here. To equal and finally surpass it in the fruits to be earned from technology was everywhere the law of its will. In short, Marxism is by origin heir to the Baconian revolution and in its own view its rightful executor, a better one (meaning: more efficient) than capitalism has been. It remains to be examined whether it can also better become its master. Our anticipated answer is that it can be so only if it reinterprets its role from bringer of consummation to preventer of disaster, that is, by renouncing its breath of life—Utopia. This would be a much changed Marxism, almost unrecognizable except for the external principle of organization. The inspiring ideal would be gone (we don't know whether the pain would be salutary or not). The classless society would then appear no longer as fulfillment of mankind's dream but very soberly as the condition for mankind's survival in the approaching era of crisis. Let us examine the chances for and against this.

3. *Weighing the Chances for Mastering the Technological Danger*

Because of the passions connected with this greatest shibboleth of our time on both sides, special circumspection is needed here. Objectivity is made easier for us by the fact that we wish to examine not the intrinsic merits of the life systems themselves but merely their respective fitness for a purpose equally foreign to both, namely, for curbing the technological impetus to which both of them wholeheartedly subscribe. The following line of argument, concerning chances which nobody can prove at this time, is in the nature of a first attempt.

a) Need-economy versus profit-economy; bureaucracy versus free enterprise To an economy governed by the profit motive, socialism can oppose the promise of a greater *rationality* in the management of the Baconian heritage. Central planning in accordance with the collective needs could avoid much of the waste attendant upon the mechanics of competition, and most of the nonsense of a market production aimed at consumer titillation. It could thus serve material well-being with greater thrift in respect of natural resources. Since prodigality in this respect is one of the ailings of the situation we are dealing with, this would constitute an important advantage of a nonprofit-oriented economic and social order. Practically, however, and according to every evidence up to now, this logically persuasive advantage is at least partially outweighed by the well-known defects of a centralized bureaucracy—misdirection from above, servility and sycophancy from below. Clearly, the logic speaking in favor of an all-powerful centralism cuts both ways: just as the right decisions at the top can prevail with greater certainty throughout the body politic and economic, so the wrong ones too must have a similarly sweeping effect; and with the stifling of initiative "from below" and the atrophying of the faculty for improvisation in the population at large, the remedies are here less at hand than in the more flexible, still comparatively open, competitive system of capitalism. In the provision of goods and services, the latter doubtlessly is doing better so far, albeit at the cost of the wastage that is becoming inadmissible: how this chosen wastefulness compares with the involuntary one peculiar to "bureaucratism" over the long run must as yet be left open in the balancing of costs. The profit-economy, to give it its due, has entries here to its credit as well as to its debit. Just as, on the one hand, it promotes waste at the consuming end by inciting wants, it has, on the other hand, an intrinsic motive to thrift at the source in the interest of lowering costs—a motive which competition then makes altogether compelling. A state economy free of competition is under no such compulsion to keep costs down in order to survive. Even a largely monopolistic capitalism, which would possess most of the detriments of a state economy without its possible social advantages, would not enjoy the same immunity: there would still remain the element of *risk* for the stockholders which affects the managers as well, and even in the absence of competition commits them to profitability and thus to economies in the production process. Even a monopoly concern can operate at a loss (and, naturally, with very different margins of profit). A civil service bureaucracy, on the contrary, has nothing to lose. The abolition of risk is a high price to pay for the elimination of the profit interest. "Profitability" thus acts as an irrational force for rationality at the production end of the economic spectrum, though for irrationality at the consumption end.

Nevertheless, we must probably grant that, taken by itself, the need standard is a better (because intrinsically more rational) *premise* for rationality than the profit standard. How this better premise is utilized is at the mercy of—theoretically intractable—psychological factors. Everything turns on *what is considered a need* (e.g., armaments: see below); and depending on how rationally or irrationally *that* is determined, on whether the present or the future governs the judgment and what perception of the latter, how assertive or temperate national egoism is in deciding policy, also how sure or unsure the regime feels itself in the favor of the people (etc., etc.)—depending on all this, the standard of "need" can cover a very wide range indeed, even to the most craven squandering of finite resources at the expense of the entire ecosystem. Certain is only that the absence of the profit motive at least removes *one* compulsion to extravagance, namely, that to the artificial *creation* of market capacities for goods neither desired before nor even known. It may also be that other compulsions or propensities toward waste are due not so much to the socialist system itself as to its provisional imperfections taken together with the pressures of its international situation. Nevertheless, the capital magnitude of the consequences of any centralistic errors when they occur (and occur they must) will persist even in the case of maximal autonomy and of a most highly improved bureaucracy, perhaps even magnified by that very circumstance.

b) The advantage of total governmental power   Admitting then, on balance, the somewhat better chances for a spirit of rationality in a socialist society, we must add to them its greater *power* to make it prevail in practice and for its sake to impose also unpopular measures. After the drawbacks of centralized power, let us now look at the advantages. These are, first, the advantages of autocracy as such, and that is what we are dealing with in the communist model of socialism (the only one worthy of discussion here). The decisions from the top, which can be made without prior assent from below, meet with no resistance (except perhaps passive) in the social body and, given a reasonable dependability of the lower echelons, are assured of implementation. That includes measures which the self-interest of those affected would not have spontaneously imposed on itself—which accordingly, when they affect the majority, would be difficult to get adopted in the democratic process. But such measures are precisely what the threatening future now demands and will increasingly demand as we go on. What we are talking of so far are the governmental advantages of any tyranny, which in our context one must hope to be a well-intentioned, well-informed tyranny possessed of the right insights. The question is then whether such an authority has a better chance to originate from the "left" than from the "right" (here in particular: whether an executive elite of that description is likely to arise from the Communist party apparatus), and this question we leave open for the moment. But since the communist tyranny already exists and from there, as it were, the first and hitherto only bid has been submitted, we can say this much: that in techniques of *power* it appears superior, for our uncomfortable purposes, to the capabilities of the capitalist-liberal-democratic complex. The real problem is this: if, as we believe, only an elite can assume, ethically and intellectually, the kind of responsibility for the future which we have postulated—how is such an elite generated and recruited, and how is it invested with the power for its exercise? At the moment we speak only of the power aspect of this dual problem.

c) *The advantage of an ascetic morality among the masses and the question of its duration under Communism*   Among the indispensable conditions of political power is the willingness of the governed to be led; and although we do not underestimate the capacity of terror to compel such willingness, that instrument is not only undesirable in itself but also over the long term of dubious success. Some identification of the community with the government, even a dictatorial one, is requisite when long-lasting sacrifices are asked. Now, a great asset of Marxism is here the emphatic "moralism" with which it pervades the society formed and ruled by it— a moralism not confined to the immediate exponents and supporters of the doctrine. To live "for the whole" and to "do without" for its sake, is a credo of public morality; and a spirit of frugality, alien to capitalist society, to which the fathers of the revolution were personally dedicated, lives on at least as a habit in the society which professes their norms. (Even lip service paid to it has its value.) In short, ascetic traits are indigenous in socialist discipline as such. This could be of great help in the impending era of harsh demands and renunciations. But here we must remember that a spirit of asceticism, certainly of frugality, was also peculiar to the beginnings of modern capitalism, and its fate there should caution us in regard to the parallel case. In the frenzy of material success it has vanished so thoroughly that its revival from within is utterly unlikely in the home areas of habitual affluence, and even its imposition from without and above will be hard to enforce. (The unpredictable intervention of a new religion must be left out of all such appraisals.) Moreover, those who hitherto promoted the abundant life are ill-suited by their own disposition to enforce its denial, and it would under their management also lack the credibility of unselfish motives which is so important for the acceptance of painful policies. It is open to question whether communism, which by itself shares the material prosperity goals with capitalism, can resist the temptation of success where it begins to taste it; whether it is willing, so long as it stands in competition with capitalism, to stay by choice behind the latter's accomplishments in the sphere of consumption; and whether, supposing it came to power in the old lands of self-indulgent affluence, it could afford to start its career there with a brusque lowering of the standard of living, especially when inflicted for the equalizing of international distribution, that is, for the benefit of foreign lands. These are hard questions to answer, and the example of Russia and China up to now tells us nothing conclusive about them. Greater frugality, to be sure, is normal there, but its being voluntary is doubtful, since inferior productive capacity can hardly be credited to a superior will. On the other hand, the life expectancy of the ascetic spirit itself need not be greater in the communist case than it was in the capitalist. But the former may be spared the test by the fact that, coming so much later, it is already close enough to the onset of globally compelled scantiness that any affluence still achieved before that has less time to corrupt it. This might enable the said spirit to pass without a great hiatus and, as it were, imperceptibly from an asceticism in the service of future wealth into an asceticism in the service of preventing future excessive poverty. China as the latest arrival in the arena has perhaps the best auspices. But one thing neither of the two communist giants has as yet demonstrated: that, ready as they are to great sacrifices for their own better future, they are also ready to make sacrifices of well-being for the benefit of other parts of the world. And that is the very thing which will be required, given the worldwide scope of the problem and the territorial inequality of natural wealth. A global "class war" of nations is taking shape, for which the supranational Marxist theory may have an answer, but regarding which the practice of Marxist national or territorial states shows no signs yet of being any freer of collective egoism, when put to the test, than other sovereignties. We shall return to this point later.

d) Can enthusiasm for utopia be transmuted into enthusiasm for austerity? (Politics and Truth)   Amid all these doubts, there remains the great asset of enthusiasm as such with which Marxism knows how to inspire its followers. Capitalism has nothing to compare with its readiness for sacrifice. There it would require a new religious mass movement to induce a voluntary break with the inculcated hedonism of the abundant life (i.e., before grim necessity coerces it). But that asset, remember, is an enthusiasm for utopia, that is, for an expected future fulfillment to be purchased with the present renunciations; and the question is how soon it would be spent when diverted to an entirely different purpose, namely, the unglamorous one of humanity tightening its belt. In any case, being used for this purpose is being abused according to its own meaning. Such a misuse, as a workable policy, would be feasible through deception (by concealing the substitution of purposes), and this would not be the first example of it in world history. What a colossal irony of fate it would be if Marxism, which has stressed so strongly the exposure and critique of "ideology," were destined to resort to this strategem of class society and to serve an altered end with a "false consciousness"—albeit this time consciously, while normally the ruling ideology is supposed to have been rather an unconscious product of entrenched interests. Here we would have the contrary: false consciousness entertained by a correct consciousness! I do not stand aghast at the thought.[3] Perhaps this dangerous game of mass deception (Plato's "noble lie") is all that politics will eventually have to offer: to give effect to the principle of fear under the mask of the principle of hope. But the strategem this time presupposes the existence of an elite with secret loyalties and secret goal conceptions, and the emergence of such a one is less probable in a doctrinaire-totalitarian society than under the conditions of independent opinion formation in the free (or individualistic) societies. There, on the other hand, its governmental power, if it attains to it at all, will be much weaker, while in the communist case a conspiracy for the good at the top, once installed there, would have all the power of absolutism plus the psychological force of the pretended ideal on its side.

We have here entered a twilight zone of politics, where the stranger hesitates to tread and rather lets the practitioners of political science try their hand: a new Machiavelli might be called for in that field—who would, however, have to propound his teaching in strict esotericism. It would naturally be better, morally and pragmatically more desirable, if one could entrust the cause of mankind to a spreading "true consciousness," able to evoke its own public idealism, which would voluntarily, generations in advance for one's own descendents *and* simultaneously for the indigent contemporaries of other nations, take upon itself the renunciations which a privileged position does not yet dictate. The possibility of this happening, given the unfathomable mystery that is "man," is not to be discounted. To hope for it is the prerogative of a faith which indeed would give the utopian "hope" a vastly different, in part more modest, in part more sublime sense. Empirically there is little ground for such a faith, but neither is there a veto against it. Responsibly to wager upon it, so it seems to me, is not possible. Enough of these speculations. It will be noted that the whole reflection had little to do with the substantive essence of Marxism and merely examined certain formal properties of its historical reality with a view to its possible service in the uncertain business of safeguarding the future. I am prepared to be accused of cynicism and will not oppose it with the assurance of my good intentions.

e) The advantage of equality for the readiness to sacrifices   There is one other point in favor of Marxism which does after all introduce its substantive essence into the calculation: real equality, if the classless society has proved itself in it, protects the renunciations that have to be imposed from the suspicion of being exacted in the interest of a privileged class. A distrust of this sort is unavoidable, and more often than not justified, in class societies, be the stratification plutocratically determined or otherwise, and it would necessitate enforcement of the needful by coercive power, where faith in the impartiality of burdens might elicit willing cooperation. *Justice,* credible in the intent and apparent in the execution, will be—even more than in the normal course of things—a *sine qua non* in the extraordinary exactions which a policy of retrenchment and conservation will bring in its wake. Feelings of inequity and one-sided victimization (even imaginary) could be fatal to the entire cause. Complaints will be numberless in any case, but they must at least admit of an answer which does not insult the moral sense. Since relations of rule exist in communist states too, a credible integrity of the party bosses (etc.) becomes here a veritably vital question (much more so than with the answerable and deposable powers in a democracy).

Now it is well-known, and could not be otherwise, human nature being what it is, that things do not look well in this respect in the communist world. The bureaucratic managers' self-endowing with special benefits from the social product is not even disguised, and corruption of one sort or another is simply inseparable from the possession of power. Regional and ethnic inequalities must be added; there is the fact of Russian predominance in the Union, and so on. Nothing in the system itself allows a prediction of whether this will get better or worse. Class privileges, even when not called this by name, are inevitable with the high differentiation of functions and thus gradation of responsibilities in a modern technocratic society—and selflessness is no more to be expected of socialist man than of people in general. As long as the status-associated, socioeconomic premiums which the managers grant themselves do not become hereditary, even striking inequalities can (in the framework of our argument) remain tolerable. On the other hand, the lack of control from below does not augur well for the restraint of eminently natural proclivities of a *de facto* ruling class. In other words, not even the "classless society" is really classless.

Nevertheless, on balance one should think that the pervasiveness of the principle of equality in any socialism is such that the latter on the whole offers a better chance for equity *and for the presumption of it* than any alternative open to choice. It would be different if democracy were among the alternatives; for when the people elect their representatives and periodically subject them to reelection, they can more or less hold them to good conduct. But in the preceding deliberations it was tacitly assumed that, in the coming severity of a politics of responsible abnegation, democracy (in which the interests of the day necessarily hold the stage) is at least temporarily unsuited, and our present comparative weighing is, reluctantly, between different forms of "tyranny." And there, socialism as official creed of the state, even when defectively practiced, and all the more, of course, when practiced nearly in accordance with the creed, can much facilitate psychologically the popular acceptance of a decreed regimen of austerity. *Provided,* therefore, that the leadership can bring itself to the right conception of things to come, which naturally is in no way guaranteed by the socialist structure as such, socialism has an advantage here, from which even an otherwise unloved Marxist system can benefit. To what extent that provision can be counted on—the great "if"—has yet to be discussed.

4. *Intermediate Result: A Plus on the Side of Marxism*

Up to now, our findings from the perspectives of social discipline as well as of social trust give Marxism an edge over other dictatorships, if one takes an instrumental view, and if one (quite hypothetically) assumes a consensus among the rulers about the coming objective priorities in the long-term cause of the world. We hinted that in this connection fictions might play a role—principally these two: the ideal of utopia, which can generate enthusiasm on behalf of overarching purpose; and the institutionally anchored principle of equality, which in the day-to-day affairs can avert the suspicion of favoritism. Naturally these are "fictions" in very different, even opposed, senses. The pretension of utopia is only useful here if it is not also the truth, that is, when at least *pro tempore* actual utopian striving is suspended. The pretension of egalitarian justice is the better the more of it is also true—that is, it must look that way, but where possible it must also be that way. (There remains, however, a political value to the appearance that exceeds reality, just as an appearance falling short of it can rob the best reality of part of its political effect.) In giving "appearance" its due, we are saying no more than that men's opinions, true or false, are a factor in the march of events. In the case of utopia, however, we are also saying that in special circumstances the useful opinion may be the false one; meaning that, if the truth is too hard to bear, then the good lie must do service. But perhaps this view underestimates "people." Perhaps the grim truth can also inspire, and not just the few, but eventually the many as well. This is the better hope in dark times.

IV. Examining the Abstract Chances in the Concrete

After the comparative, instrumental analysis, which—so it seems— leaves an inwardly sobered "Marxism" with a plus (namely, the better chances for being able to meet the harsh demands of the future), the main question arises: What are the chances that "it" will also avail itself of its better chances? This question requires, as it were, a metacritique of the preceding critique and thereby lays the already concluded comparison once more open to review. For "Marxism" is naturally an abstraction, and what we are concerned with are concrete Marxist regimes and concrete communist parties. We thus pose the, only apparently reiterative, question concerning the chances of chances.

1. *Profit Motive and Stimuli to Maximization in the*
   Communist Nation-State
Among the first order chances, we named (pp. 145–146) the rational superiority of the need-standard over the profit-standard. However, we must now ask: Is the profit motive actually eliminated in a communist society *per se*? For the private (individual) sector it is indeed, or mostly, since the system offers it no opportunities. But what precludes a *collective* striving after profit, namely, at the expense of other parts of the world? What, given sufficient power, precludes an economic imperialism which in ruthless exploitation of the natural resources and economic potentials of foreign nations (possibly with the aid of local "communisms") need not fall behind capitalist colonialism? One will rejoin: Communist ideology stands in the way. But there is surely no relying on that. For we have come to know too well the combination of "socialism in one country" with patriotism, collective egoism and national power politics. At least the nonsocialist world can be treated as an object, if not as an enemy. And still less can better insight into planetary necessities be counted on. For, first, we are inquiring precisely into the *chances* for such insight to exist in Marxist systems of rule; and, second, we have already before us evidence of the fact that there, too, correct insight can be limited to one's own territory. At the Bucharest Population Conference in 1974, China, which at home appears successfully to control the growth of its population (in itself a shining example of what a communist regime can accomplish), disgracefully and against its better knowledge advised the "Third World" not to let itself be persuaded by the "imperialists" to do the same—out of a cynical interest to hinder a premature pacification of global tensions that might redound to the enemy's advantage.[4] This, admittedly, is not a case of economic exploitation but one in terms of power-politics—in our context (considered ecologically) perhaps even more pernicious than the former. But to stay with economics, the maximization motive, which *in its effect* we equate with the profit motive, is as congenital with Marxism as it is with capitalism, and Marxism too might find it difficult to check for internal as well as external reasons. The internal reasons are ultimately bound up with (materialistic) utopianism, which will occupy us separately. For now we only repeat what we mentioned once before—that the principle of need still leaves open what will count as needs and in what order

of priority. Something of this can be gauged already on inspecting the external grounds for maximization. "Bad examples corrupt good morals": Coexistence with the successful industrial systems of capitalism almost compels Marxist leaders to offer their own people something comparable. And if (as the master plan foresees) they are swallowed up in the expanding communist body, the contagion from without becomes one from within. Meanwhile, the coexistence engenders, *inter alia,* the compulsion to arm, which can grow to extremes of economic extravagance: the more so as material welfare, if only for the sake of internal stability, must not be too much forgotten over this. Thus internal and external temptation join hands, and industrial growth must forge ahead. But ravagers of the earth are *all* modern industrial societies, and hence today Soviet Russia is also.

2. World Communism Is Not Immune against Regional Economic Egoism

Therefore, it will be said, "socialism in one country" (or in a few) is at best a first step and only a socialist world order can be the solution. But two things are crystal clear in this prospect. First, that condition could, as things are, only be brought about by a world war of truly global scale, after which most of what occupies us here would no longer be there to worry about. The object of all our theoretical effort here is how to avoid a gradual catastrophe for humanity, and to replace it with an abrupt one would be an absurd solution indeed. Second, even if the establishment of this socialist world state could go off more mercifully, it would still be a foreign rule for many of its parts—for the rich, if they are forced to lower their standard of living, for the poor if they are barred from rising to it—and regional freedom movements, with all the passions of nationalism, would be inevitable. This is to say that a centrally governed world state, which in its necessary partiality from regional points of view will be felt there as an oppressive foreign rule, will be unstable, and its socialism would make no difference to that. Nor would it make any in a federation of sovereign member states. Why should there be no conflicts—or even war—between socialist republics over finite vital objects? Collective entities will continue to be territorially and nationally constituted; of collective altruism we have no record in world history and not the slightest grounds for expecting it in the future. As long as the general trend is upwards, that is, toward continuous economic growth, and this keeps, even slightly, ahead of population growth, it will in principle be possible to assure all subdivisions of *relative* satisfaction, as there will be enough to let each have its slice of the increment and even the poorest may feel an improvement and can hope for more. Under such conditions, a world union of Soviet republics can persist even in the midst of great regional inequalities. In other words, successful (namely, peaceful) world socialism is tied to economic *expansion,* without which the destitution of vast regions of the earth cannot be alleviated—unless it be through a radical international redistribution of existing wealth (and this, as noted, would require that use of force which would not leave much to distribute). But economic ascent on a worldwide scale will soon be a thing of the past, and although regional improvements are still possible (and imperatively necessary), the total process tends to slow to a standstill, and in the advantaged regions of the world even to contract. This global direction of things lies beyond the difference between socialism and capitalism. A world socialism will inherit the division of the world into "haves" and "have nots" (Russia already belongs to the former, China perhaps soon will), and along with the inequality it will inherit the respective resistances against equalization: these will simply cease to be foreign and become "domestic." Separatisms of every sort would spring up predictably and could be pacified only by special concessions at the expense of the general policy of retrenchment.

3. *The Cult of Technology in Marxism*

To this must now be added the *constitutional,* inner reasons which make the departure from economic maximization especially difficult for Marxism, and this already on the single-nation scale prior to any international application. First on the list is the cult of technology, which here enjoys a force of faith unknown to this extent in the West. We said earlier that Marxism is a fruit of Baconianism and in its basic self-understanding regards itself as its chosen executor. From the beginning it has celebrated the power of technology, from whose marriage with socialization salvation will spring. Taming it is not the point but freeing it from the bonds of capitalist ownership and putting it, thus liberated, as itself a liberating force in the service of human happiness. The West, which invented the whole thing, is more skeptical here. At the word "alienation" *we* think involuntarily of the worker's alienation from the object of his work by the machine and from the sense of his work by the fragmentation of the production process into monotonously repeated, "soulless" manual suboperations; and beyond this we may think of the alienation from nature by the thoroughly artificial world of the big city. But the Marxist, quite to the contrary, speaks of the "humanizing" of the world by nature-transforming human labor (nothing is farther from his mind than a sentimental or "romantic" view of nature). And by "alienation," if I am not mistaken, Marxist literature does not mean alienation from the process and object of labor through the machine, but rather alienation of the maker from his product by the alien ownership of the means of production (and thereby of the product as well). *This* "alienation" is remedied by the workers' ownership of the means of production and of the product of his labor, thus by socialization, which rather intensifies "technological alienation" still further by encouraging the utmost of rationalization. Misgivings in this regard, any resistance against the "deadening" of the work process entailed by it, would be dismissed by the orthodox Marxist as reactionary romanticism. But what truly goes beyond the bourgeois liberal attitude is the almost religious faith in the omnipotence of technology to the good. Those old enough to have seen the beginnings of the Soviet venture will recall such slogans as "Socialism Is Electrification," book titles like *Concrete,* Eisenstein's heroizing film on a railroad construction, the glorification of tractors, the celebration of every large new manufacturing plant, of every engineering advance, as contributions to socialism. One may have an understanding smile for this childlike early phase. But much later, and not at all childlike, Malthusianism was officially "condemned" as a bourgeois class doctrine, and—long before China—Moscow had proclaimed that a socialist-led science and technology in food production could keep pace with any population growth. Indeed *the very idea of a natural limit* to the *exactions* of human skill was expressly repudiated (not to mention limits to the advancement of that skill itself; more on both aspects in Chap. 6). Scarcity is the fault of either insufficient technology or malicious market manipulation, and even the former can in the long run only be the fault of the class system. It makes little difference how much of this is honest conviction and how much hypocrisy, since the latter is summoned to please official doctrine and is made honest, as it were, by being made binding.

Another example of the rule of ideology over what is "truth" is the famous Lysenko episode in the quarrel about party-correct genetics. Over the immediately grotesque aspect one must not overlook what lay at the bottom of it: a principled commitment to a *technological* conception of *Society.* In the role of "biological environment," society must have the power to form man even biologically, hence via its own transformation also to *trans*form him in adaptive conformity with itself. Since its own transformation, however, comes about by design and plan, the manipulation of societal *conditions* becomes the artful means for the hereditary manipulation of man, and society as a whole becomes an engineering technique for the determination of his nature. Of the, then still new, "Western" genetics, Moscow perceived at the time only the fatalistic, thus "reactionary," aspects (immunity of the genes to environmental influences, with implications for race theory, etc.) and not the also implied manipulative, thus "progressive," possibilities—which, it is true, would set in at the individual end and give room for individualistic arbitrariness. According to Marxist doctrine, determinism must pass through the collective—which shifts arbitrariness back one step, from the individual to the willfull shaping of this determinative collective itself, namely, society. Actually, then, Marxism can easily make its peace, over the corpse of the Lysenko error, with the engineering potential of Western scientific genetics, if only this potential comes under the control of "the society."

In using it, a Marxist authority will by its nature have fewer inhibitions than the bourgeois Western world, where still so many, various remnants of tradition and religion can interfere with choices. That "opiate of the masses," which religion was once said to be, has become in the meanwhile technological progress, and it is to be feared that under Marxism it is this even less than under capitalism *only* for the masses. The important point here is that the technological impulse is built into the essential nature of Marxism, and to resist it will be all the more difficult as it is bound up there with a stance of extreme anthropocentrism, to which all of nature (including human) is but a means for the self-making of still unfinished man. This psychological factor, though not quantifiable, must be counted in the evaluation of chances in which we are engaged.

4. *The Seduction of Utopia in Marxism*

The greatest temptation from within lies in the inmost soul of Marxism— "utopia." It is its noblest and hence most dangerous temptation. For, of course, it is not the case that the "materialism" with which it explains history, or even its simplistic materialist ontology, portends a banal-materialistic content of its *ideal,* namely, an ideal of the full stomach. The materialism applies to the conditions, not to the goal. In "First comes the feed, and then morality" (Bertold Brecht), both the "first" and the "then" must be taken seriously. It purports that he who starves, no different from one who is suffocating, and generally he who lacks the first necessities of life, is locked in a premoral state and excused from further demands; but that "then" indeed "comes morality" and makes its claims upon him. Thus the insolent banality from the *Three Penny Opera,* which apparently trivializes the "morality" thus postponed into a luxury of those who can afford it, in fact stipulates a moral duty for third parties, namely, to help overcome that state which prevents morality and replace it with one in which it can no longer be said, "But circumstances won't permit, alas." If this were the whole utopia of socialism, who could take exception—aside from perhaps doubting the choice of path? But with such an interpretation in terms of charity, Marxism at least would be vastly underrated, just as it would be with a purely economic interpretation of the altered distribution of the social product. "Justice" would come closer to the truth, but even this describes less the ideal itself than a precondition for it: the more just order of things is meant to open the gates for the "authentic" human potential waiting to be set free, which has hitherto been obstructed by unjust conditions on *both* sides of the class divide. That is the meaning of the otherwise enigmatic contention that all previous history, including the present which still antecedes the classless condition, was merely "prehistory," and that the true history of mankind will only begin with the new society. Wisely, nothing is said about *wherein,* apart from the removal of earlier wrongs and conflicts, the higher realization of "Man" is *positively* to consist: and rightly so, for with only past, inauthentic history to draw from, the Marxist must not profess to have of this a substantive anticipation. But *we*, confronted with the magic-empty call of utopia, summoned by its promise, are entitled to ask, *what* the more just condition—incontestably a value in itself, for which one might well be ready to pay in other human coin, such as cultural brilliance—might beyond itself bring to light (or allow to come to light) of man's nature that was previously repressed; and what supports the belief that such a thing exists and only awaits its redemption.

V. The Utopia of the Coming "True Man"

1. Nietzsche's "Superman" as the True Man to Come

Since on these questions utopianism itself is silent, one might as well turn to Nietzsche, who from the opposite end of the spectrum arrives likewise at the belief that everything past was only a preliminary and man himself is only a transition from the animal to the coming superman. He more than the Marxist allows us a glimpse of his idea of human greatness which casts its shadow ahead into the future. Nietzsche notoriously was no utopian, a coming general condition held no promise for him, and least of all a "definitive" one, which to him would spell the death of the human experiment. To be but a transition applies strictly to man as such, thus also to the expected "superman" himself, who again would aspire beyond himself, and so on into an endlessly open horizon. We also know that Nietzsche had nothing but contempt for the blessings of socialist equality and for any general happiness, and thus can hardly be considered a supplementary source for filling in the contents of the utopia left open by Marxism. Nevertheless, with his lively sense for the humanly great and small, sublime and pitiable, hale and distorted, he halfway knew of what he spoke when he expected the greatest still to come. And there one need only look at his heroes (most of them also in some sense his enemies): one will recall the "dialogue among the peaks," which the great minds conduct across the lowlands of the ages and will presumably continue to do in the future. Will the peaks then be higher still? It is hard to imagine, indeed, utterly opaque, what that can mean. Perhaps they will be less rare, more densely spaced? Possibly; but one does not quite see that numbers were of interest to Nietzsche here. And should we impute to him the thought, "What might a Plato not have been if only he had not been led astray by his two-worlds theory? What a Spinoza, had not his Jewish hatred choked on the Jewish God? (etc., etc.)"? Impossible. He would not have wished any of his peaks different. Admittedly, they were not yet the entirely "free spirits" of the future, of whom, after the death of God, the final loss of transcendence, a previously unknown "hardness" and "bravery" will be demanded. That is the only distinguishing virtue of the superman Nietzsche can name,[5] and this is truly making a virtue of necessity. Besides, even this distinction is only an enhancement of what he admiringly already knew from the past. In short, "the superman" has of old been with us, just as "man" was; and although future superman will surely be different from all his predecessors, so was each of them in his time. Finally, there is not a hint in Nietzsche concerning what can be done concretely to bring about the higher man, to enable and expedite his appearing, or even merely the probability of it (unless one takes occasional similes from the sphere of breeding for such hints).

2. *The Classless Society as Condition of the Coming True Man*

In this, Marxism has the advantage of a political concept and of an action program over the hope of the visionary: it knows the way to the *conditions* for the higher man. The way is revolution, and the sum of the conditions is the classless society. Behind this lies a premise which separates Marxism from Nietzsche (and from the majority of the classical philosophers) and which it shares with most of the credos of progress: man is basically "good" and only made bad by circumstances. All that is therefore needed for his essential goodness to become actual are the proper circumstances. Or, what comes practically to the same, man—with *no* determinate essence—is the product of circumstances, and good circumstances will produce good men. According to both views, the goodness or badness of men is a function of good or bad circumstances. At least one aspect of progress is always the clearing away of obstacles. Viewed thus, according to Marxism, the circumstances up to now—namely, class society and class struggle—were never yet good and so neither was man. Ergo, only the classless society will bring with it the good man. This is the "utopia" in the essence of Marxism. Now, "good" can mean two things here: goodness of character and behavior, thus *moral* quality; and productivity in supraeconomic values (for the economic values are just conditions) and their goodness, thus *cultural* quality. In regard to both qualities if possible and to one of them at least, the classless society is held *ex hypothesi* to be better than any before it and indeed to bring man's true potential to light for the first time. What can this mean?

a) Cultural superiority of the classless society?   Regarding the "cultural" aspect, we are back at the quandary of Nietzsche's riddle: Will there be greater geniuses? More of them? Happier ones? More beneficial to society? We know utterly nothing of the conditions conducive thereto. Concerning the "more" in numbers, it is possible that many a talent otherwise held back by poverty would come to its blossoming, and that would be a gain. On the other hand, much talent might also be suppressed by the heightened societal censorship, and the net balance is not predictable. Still less is the quality. About the riddle of "genius" we better keep silent anyway: usefulness is probably the last thing that should be imputed to him. Individuals aside then, can we in terms of overall level expect a greater art from this society than any of our dark prehistory has left us? A better-behaved perhaps. A yet mightier science? One more narrowly geared to public utility perhaps. Even the "perhaps" is said ony very offhand. In truth, we again know next to nothing of the conditions for creativity, collective or individual, nor anything predictively useful on how and where, in which epoch and society, a great art (e.g.) will come forth rather than a mediocre one. Only of this can we generally be sure that mediocrity is more the rule—in the classless society as much as in any other. It may then make a difference where the greater rewards go, whether to the exceptions or to the conformist rule. Even there, the effect is not certain so long as the exceptional is not too oppressively penalized. But another matter is that we might well be *willing* to pay the price of cultural brilliance and originality for a social order which is more just and less disfigured by human misery, if those cultural qualities were tied to such negative conditions. One could think it *right* that for the sake of decency we put up with the prospect of a pervading philistinism: and to say Yes or No to *that* should be the moral stance toward the whole problem, instead of the childish wish to have everything (= Utopia). But if one gives the downright ethical aspect precedence in this (perhaps demanded) choice, then one must well distinguish between its immediate imperatives, which are valid in themselves, and the in turn ethical hopes attached to their fulfillment—hopes this time "utopian" in terms of morality itself. Precisely such hopes—namely, for a morally "better" man in general—are the core of the ideal for which one would have paid the aforementioned price in the crust. How do matters stand in this regard?

b) Moral superiority of the citizens of a classless society? Here is where the doctrine of "goodness" as a function of circumstances comes preeminently into play. What plausibility it has points much less in the direction of cultural creativity than in that of the *moral* condition of the community. There indeed it will be readily granted that with a just, less unequal distribution of vital goods, as a socialized economy is intended and expected to bring about, many of the incentives to violence, cruelty, envy, avarice, fraud, and the like will disappear, and perhaps a generally more peaceable, if not more brotherly, disposition will prevail among people than in the pitiless bustle of competition, where "the devil takes the hindmost." Evidently, crimes and vices born of want must diminish when the want diminishes. Less evident is what other motives or occasions for badness will take their place. (One thinks immediately of political ambition.) For them not to be lacking, human nature with its susceptibility or its ingenuity can be relied upon. That should not deter from the removal of those motives, pressures, and opportunities that can be removed and whose existence is a scandal in itself. But whatever the net balance of such cost/benefit accountings may be, no knowledgeable person can seriously believe that, with a certain set of contrary stimuli removed, people everywhere will become good-natured, unenvious, fair, brotherly, even loving toward each other to a hitherto unknown degree—that, to put it differently, they in a body will not just willy-nilly obey, but also internalize the institutionally incarnated, so to speak "objective" ethics, and then spontaneously practice it; or, that the state will ever consist of virtuous citizens only. (For more about this, see Chap. 6.) The socialist state itself does not believe it, as the thoroughness of its police and informer system proves. "Socialist man," in some respects perhaps better placed morally than he who fends for himself, can still by any number of criteria be either a good or a bad man. That certain social excuses can no longer be adduced and guilt thus becomes more unequivocal, can ethically only be an advantage. But as long as there is temptation—and the human heart will always see to this (as we should hope rather than fear)—there will always assert itself the fact that men are men and not angels. One is almost ashamed to state this truism. Why are we doing it?

c) Material prosperity as a causal condition for Marxist utopia   We do it because of the dangerous power of *one* temptation: utopia itself! Its more general dangers, for example, the well-proven temptation to pitiless policies in its pursuit, are not our concern now. Its special danger in the context of our theme consists in the fact that among its causal conditions it must avoid poverty and seek, if not abundance, at least a satisfactory plenitude of physical existence. The materialism of its ontological hypothesis renders material well-being an imperative presupposition for the goal of setting free the true human potential: though not itself the end, that well-being is the indispensable means to the end. Thus the pursuit of plenty with the help of technology is elevated above the grosser and all-too-human incentives shared with capitalism, which anyway work in this direction unassisted by ideology, and obtains the rank of sublime duty for the servant of utopia: nothing less than making possible the advent of true man demands it. About this we have to say two things which nobody says gladly: first, that we today cannot afford a utopia bound to this condition; and second, that also in itself it is a false ideal.

VI. Utopia and the Idea of Progress

1. *Necessity of a Farewell to the Utopian Dream*
a) The psychological danger of the prosperity promise   That we can no longer afford much heightening of prosperity *on the global average* has been argued earlier on grounds of environmental and resource tolerance. The latter's room for play is largely already preempted by one overexploit-

ative segment of the whole. But we surely can make that average more even. This means self-denial for the developed countries, as raising the standards for the undeveloped is (without reckless violation of those boundaries) only possible at the former's expense, namely, by regional shifts of productive and consumptive capacities away from their densest clustering at one extremity of the spectrum: the closing of the gap—any tangible converging thereto—will have to fall somewhere *between* the present, yawning extremes. Even there, granting a lot of "give" from the privileged side, the margin of attainable (and sustainable) gains for the immensely larger, needy side is modest. That is to say, the "somewhere" of ideally *possible* equalization (or realistically probable approximation) lies much closer to their than to our, overendowed, end of the scale. Even the most ruthless redistribution of already existing global wealth, or rather of the industrial capacities dedicated to its production (which, however, could not be achieved peacefully), would barely suffice for that raising of the living standard of the impoverished parts of the world which would just do away with misery. More likely, we could count ourselves fortunate if it would merely halt its further growth with the concurrent, ungovernable population growths in those same parts. And even in the best case, that is, with any partial success, it can only whet the appetite for the (still inadequate) "more." Nevertheless, it is evident that something *must* be done in this direction, although it will necessarily lag far behind any utopia-promoting abundance. It is then just as evident that the prosperity of America, for example, which consumes much too great a share of the world's treasures for its prodigal life-style, must suffer palpable losses, whether voluntarily or coerced by the "class struggle" of the nations (the same goes for Europe, et al.), and this is, at least psychologically, anything but a recommendation for utopia. On the contrary, utopia must commend itself with promises of broader mass appeal than that of equal justice for all. The majority must foresee itself as a winner, at least not a loser, and particularly, of course, the majorities in the powerful, that is, the wealthy nations, upon whose cooperation all of this will depend—but they are precisely the ones who will have to foot the bill. Here, therefore, the spell of utopia, with its promise of better things to come, can only prove obstructive of what really needs to be done, since it directs action toward the "more" and not the "less." Warning rather than promise—against evil things to come—would be not only the truer but in the long run probably also the more effective politics.

To sum up: the watchword will have to be contraction rather than growth, and this will be harder to adopt for the preachers of utopia than for the ideologically unencumbered pragmatists. So much for the danger of utopian thought in this hour of world history—a contribution to the appraisal of "chances" which we are conducting with regard to Marxism as a pretender to the world's cause. Already from this purely pragmatic reckoning there ensues the sober command of maturity to bid farewell to a cherished dream of youth—which the hope of utopia is for humankind.

b) Truth or falsehood of the ideal and the task of responsible politics   The time for this has also come in a more intrinsic, noninstrumental respect. It is time to ask what the dream of utopia is really worth; what would be lost, or what perhaps won, with its abandonment. If the dream is untrue, then at least truth would be gained with abandoning it, or at least a greater adjustment to truth. But truth, or even mere proximity to it (which here would consist in the negative mode of a skeptical withholding of faith) is not always salutary. The psychological value of utopia—that it can inspire great masses to deeds and sufferings to which they would otherwise not bring themselves—is incontestable. As a historical force, "myth," whether true or false, was often indispensable (at that, by the way, for good as well as for evil). Such a myth is utopia, and it has worked wonders. But here, too, a maturity is conceivable which can do without the deception and, for the mere preservation of humanity, takes sacrifices upon itself which hitherto only the luster of promise could induce: thus, out of selfless fear instead of selfless hope. Most certainly must this be so among the leaders, lest *they* succumb to the lure of their own ideal when matters demand to do in earnest without it. Thus at least for them the question of truth plays a decisive role, and this beyond the empirical question of the ideal's attainability, sooner or later, whole or partial, at a stroke or in stages, probable or improbable: for them, the first question must concern the "truth" of the ideal in itself, its *intrinsic rightness* and thus its legitimate positing as a goal. For only the genuine farewell to the conception as such, not a mere adjournment for pragmatic reasons, gives those who must lead into the future the moral-intellectual freedom which they will so sorely need for the decisions awaiting them. What they then do regarding the public communication of the sobering truth belongs to the art of politics and not to philosophy.

Let us therefore raise the question of truth and concentrate upon the *moral* promise of utopia, to which the—so much vaguer—cultural promise is only added as an ornament, and which retains its power as a spectral goal even when the *material* utopia, supposed to be its precondition, is recognized as, for the time being, unattainable. Is there sense in saying that "Man" as a species becomes morally better and wiser? Does the concept of *progress* apply here?

2. *The Western Idea of Progress*

The word "Progress" looms large in Western speech and sentiment, and spans a multitude of referential uses. It is applied to things social, political, cultural, intellectual, technical, and moral. Progress with a capital "P" refers to the public rather than the private sphere, and this already is a peculiarly "Western" fact. While there is hardly a civilization any-where and at any time which does not, or did not, speak of individual progress on paths of personal improvement, for example, in wisdom and virtue, it seems to be a special trait of modern Western man to think of progress preeminently as an attribute—actual or potential—of the collec-tive-public realm: which means endowing this macrodimension in its trans-generational continuity with the capacity, the disposition, even the inbuilt destination to be the substratum of that form of change we call progress. Thus employed, the term tends to be descriptive and prescriptive at once: on the one hand, describing an observable trend of serial change, as when we note the progress of urbanization or bureaucratization in past and present; on the other hand, prescribing a choice, as when people or parties call themselves "progressive." The connection is intriguing: with the *judg-ment* that the general sense of past change was upward and toward net improvement, there goes the *faith* that this direction is inherent in the dynamics of the process, thus bound to persist in the future—and at the same time a *commitment* to this same persistence, to promoting it as a goal of human endeavor. Progress is a law as well as an ideal. This com-bination of a *perceived* vector property in the movement of history as progress-bound with *adopting* this vector into the conscious will, which thereby makes itself its further agent, is perhaps the most uniquely West-ern trait. In the way of self-fulfilling prophecies, it makes to some extent its view of history come true by being a historical force itself. Let us take things from there.

3. *Technology as a Vehicle of Progress*

In our time, technology has become the dominant symbol of progress, at least its most visible external measure. In that connection, progress comes almost to be equated with material betterment. Advancing tech-nology is expected to raise the material well-being of mankind by height-ening the productivity of the global economy, multiplying the kinds as well as the quantities of goods which contribute to the enjoyment of life, at the same time lightening the burden of labor. How much more of this, on a worldwide scale, can still be squeezed from the natural environment without lasting injury to it is a question apart. And a mere consumer dream, whether vulgar or refined, is questionable in itself. But, of course, there is more to the idea of progress than a "better life" in terms of greater consumption of greater varieties of goods at greater ease: a better mankind is expected from it in terms of ethical and cultural quality and of socio-political order. It is this nobler faith which elevates all progressivist or utopian persuasions above their materialistic underpinnings. And because today this faith in collective human improvement, or in the perfectibility of "Man" writ large, enlists technology as its vehicle and thereby adorns the latter with the halo of a mission which already Francis Bacon had assigned to it, we must cast a critical look at the possibilities and limits of progress in general, lest we concede too much to the demands of its alleged vehicle. We repeat our question: Is it meaningful to speak of man in the generic sense becoming morally better or wiser than he always was? Is the concept of uniform progress applicable to that class of communal attributes? Let us examine the problem of "moral progress."

4. *Concerning the Problem of "Moral Progress"*

One can hear it often said today that progress, so far, has been lopsided in that moral progress has not kept pace with intellectual, that is, scientific-technological, progress; and that, within the intellectual orbit itself, the knowledge of man, society, and history has fallen behind that of nature; and that both these gaps must be closed by a respective catching up in the backward areas, so that man catches up as it were with himself and straightens out his hitherto uneven progress, during whose one-sided pursuit he had neglected the other side. This fashionable view expresses, I believe, a thorough misapprehension of the human and, especially, the ethical phenomenon. The fact that we have not "yet" a knowledge of man, society, and history comparable to that enjoyed in natural science has the simple reason that they just *are* not "knowable" in the same sense as "nature" is, and that what *is* similarly knowable in them is rather peripheral to their real knowledge. More than that, *these* "knowables" of the human field fit only too well into the system of the "value-free," scientific-manipulative knowledge of nature; their knowledge itself lies on the far side of the gap and thus does nothing to close it. This must here stand as a summary assertion for which we can only appeal to the thoughtful to reflect on the different senses of "knowing" relative to different kinds of objects. "Knowledge" of things human, called understanding, means something different in kind, less incremental and cumulative than that of the physical world, rather something each generation must recast in a way unique to itself, or to one or several of its rival camps. Factual knowledge, to be sure, may increase in the humanities too, but the "truth" about man which it conveys reflects more what he happens to be at the time of looking at himself than what he is once for all. Again, what he happens to be at the time, he is (in part) thanks to the way he has come to look at himself. In other words, the knowledge of things human, of the subject of history, is a historical phenomenon itself, as it springs from the shifting self-interpretation of the changing historical subject. The model of progressively superior theories, each a better approximation than its predecessors to an always selfsame "given," does not apply here. To elaborate on this would lead us too far afield. Suffice the statement: What we here have, compared with the univocally "progressing" (increasing) knowledge of one constant reality in the natural sciences, is not a cognitive lag but a cognitive heterogeneity which no progress will bridge. While a generation has the sciences it inherits (with or without its own addition), each generation has the humanities it deserves.

But what about *moral* progress, whose desirability nobody will deny, and whose possibility few would wish to deny? Does *it* occur as a collective, transgenerational movement with enduring, incremental gains, similar to the progress in science and technology?

a) Progress in the individual   There is irony in the fact that the very idea of "progress," nowadays so largely appropriated by the public sphere, has its origin here, in the moral and personal sphere, and this with an entirely individualistic reference. Bunyan's *Pilgrim's Progress* treats of the soul's progress toward salvation; and already from Socrates's time it is accepted truth that virtue grows by virtue and is the cumulative product of constant *education,* in which right company, example, practice, understanding, and, foremost, continual striving play their part. The ultimate force for progress here is "love of the good," first awakened from without by way of emulation, then more and more made one's authentic own. With or without philosophy, there was never a doubt that the *individual* is capable of improvement (one does, after all, make progress in school and also in bodily skills), nor that there are aids for it in the human environment and pathways in the soul itself with stage succeeding stage— thus also a possible movement of (possibly interminable) progress toward the better in many ways. Indeed, since every life starts out with nothing and must acquire everything, progress in *some* ways is the necessary law of development already for the *becoming* of the person, in which therefore everyone must at least *have* partaken at one time; and the question then is merely whether or not this becoming toward the "better" goes on beyond youthful learning and into biological maturity. Here, ethical doctrine has always held that there ought to be no end of it until death, neither in knowledge nor in abilities nor in moral character, that education should continue as self-education throughout maturity—and that it can do so because the more perfect and perhaps attainable lies always still beyond the attained. *Here,* therefore, the idea of progress has its origin, as a concept and as an ideal, and there is even room for a personal "utopia." But all this refers to the single person, the psychophysical individual, and most of all to his "soul." Does something similar hold for the collective? For the group, for historical societies, perhaps for mankind as a whole? Is there such a thing as a "moral education of the human race"? Put most generally: Is what holds for ontogeny transferable to phylogeny?[6] We have already discussed this question once (Chap. 4, IV.2 f.), namely, in the context of historical prediction and of the knowable future-horizon of responsibility; and we have found that in most respects there is *no* valid analogy between individual lives and those of historical groups. Nevertheless, the concept of progress has a place in both, though with different senses, and we will now consider the phenomenon of historical progress.

b) *Progress in civilization*   Without doubt, there is progress in *civilization,* generally in all those modes of human ability that are capable of increment beyond individual lives because they are transmissible and a collective possession. Most obviously, this applies to science and technology. Progress is also a fact, if not quite so unequivocal, in social-economic-political order, in security and comfort of life, in the provisioning of needs, variety of culturally generated aims and modes of enjoyment, broadening of their accessibility, in the rule and quality of law, in public respect for personal dignity—and, of course, also in the "manners," that is, in the exterior and interior habits of living together, which can be more crude or refined, harsher or gentler, more violent or peaceable (which may even result in the formation of "national temperaments"). In all these matters, there can be progress to the better, at least to the more desirable—and, as we know only too well, also again regression, sometimes of the most horrible kind. But on the whole, it seems, we are allowed to speak of an "ascent" of mankind until now and also of further prospects for it in the future. Only, as we have learned by now, a price is to be paid for it; with every gain something valuable is also lost; the human costs of civilization are high and, with further progress, still bound to rise—that hardly needs elaborating anymore. Nonetheless, even if we had the choice (which for the most part we have not), we would be willing to bear those costs or let "mankind" bear them—except such as would deprive the whole enterprise of its meaning or even threaten to wreck it. Let us, then, review what "progress" specifically means in the different areas just enumerated.

5. *Progress in Science and Technology*
    Clearest is the case of natural science and technology. Here, constant accretion is not only conceivable according to their nature, but in fact, albeit with interruptions, has also taken place throughout human history to this day in the most visible and indisputable manner. And everything in the present state of both these twin enterprises points to indefinite continuation of their movement in the future (perhaps even with exponential increase in their output). At any rate, they are in principle capable of it by both their own cognitive formalism and the inexhaustible nature of their objects, and no limit is set to them from either. Here, then, progress, even potentially endless progress, is an unequivocal datum, and its "ascent" quality—meaning that the later is always superior—is by no means a mere matter of interpretation. Less clear is the question of cost.

a) *Scientific progress and its price*   As regards *science,* unendingness of its task and thus of its possibilities is rooted in the nature of its object (the physical world) and of cognition itself, and the pursuit of this open-ended road is more than a basic right, it is a high duty of the cognitive subject endowed with the capacity thereto. This cognitive subject, however, is ever less the individual mind and evermore the "collective mind" of the society that stores the knowledge; and therein lies the inward *price* to be paid for scientific progress, namely, the price in the personal quality of the knowledge itself. Its name is "specialization." Enforced by the monstrous growth of the contents of knowledge, their proliferating subdivisions, and the differentiated, evermore subtle methods developed for them, it leads to extreme fragmentation of the "existing" total of knowledge among its adepts. The individual pays for creative collaboration in the process, even for adequate comprehension as a spectator, with renouncing co-ownership in everything outside his narrow competence. Thus, while the total stock of knowledge grows, the individual's share shrinks to ever-tinier portions of it. We are speaking of the participants in the scientific process, the researchers and experts themselves. Outside this charmed circle, the whole of knowledge and each of its parts become evermore esoteric, ever less communicable to laymen; and thus the scientific estate excludes the overwhelming majority even of literate contemporaries from the role of intelligent witnesses. Genuine knowledge of nature may always have been the preserve of a small elite, but I doubt that the educated contemporary of Newton stood so helpless before his work as his present counterpart stands before the mysteries of quantum mechanics. The chasm widens, and in the ensuing vacuum pseudosciences and superstitions spread. Still, nobody will plead in earnest to halt the process. Going on with the adventure of knowledge is an august duty, and if such is its price, it has to be paid. Here, then, is an indisputable case, perhaps the only entirely indisputable one, where possibility of unceasing progress coincides with its desirability, even with an imperious claim to our affirmation. But of course, being always and forever unfinished, that pilgrimage of progress has nothing to do with bringing on perfected utopia. Its victories and defeats in the remote regions of theory neither help nor hinder utopia's coming. Nor is, conversely, as the historical record shows, utopia or its expectation needed for the vigor of the theoretical urge and its harvest of progressive success. The best we can hope for from utopia in this respect is that, once in power, it will forbear interfering with and inhibiting the life of theory. But utopia or not, what matters is that the truth as such and, almost more so, the striving for it, grace by their presence any state of man, just as their disappearance would disgrace it.

b) Technological progress and its ethical ambivalence Things are different with the robust offspring of natural science, *technology*. Since it changes the world and profoundly affects the conditions and forms of human life, to some extent even the condition of nature, technology may well have a bearing on the coming of utopia as also on its projected contents. As a matter of fact, the various utopianisms themselves, political and literary (barring the "arcadian," which I cannot take seriously), quite emphatically enlist technology as a major means to their ends, if they are not technological first and last. Conversely one may, depending on one's outlook, either hope or fear from utopia that it will speed up, slow down, or altogether stop technological advance. It is also entirely meaningful to examine (as we have done) capitalist and socialist models of society as to their comparative aptitude and likelihood to stem the technological tide—which implies that such a stemming can become desirable. That possibility must be admitted because technology, unlike science, justifies itself only by its effects, not by itself, so that, given certain effects, further advance can indeed become undesirable. Still, technology shares with its begetter-become-twin, that is, with science, the notable property that "progress" as such is an objective phenomenon inherent in its autonomous motion, in the sense that every "next" is necessarily superior to its "before." One should note that this is a descriptive statement and not a value judgment. The greater destructive force of the latest atom bomb may be deplored as adverse to value, but what is deplored is precisely its being technically "better" and in this sense, regrettably, a progress. The equivocal character of technological progress goes perfectly well with its unequivocal fact. Thus, if anywhere, then in science and technology, especially since their intimate symbiosis, we have a unique case of a "success story," and a constant one at that, grounded in their internal logic and therefore promising the same in the future. I do not think that something comparable can be said of any other time-spanning, collective human endeavor. In the case of technology, as pointed out before (Chap. 1, IV.1), this success, with its dazzling public visibility in all spheres of life—a veritable triumphal procession—brings it about that in the general consciousness the Promethean enterprise as such moves from the role of mere means (which every technique is by itself) to that of the end, and "the conquest of nature" appears as the vocation of mankind. *Homo faber* towers over *homo sapiens,* whose knowing is made a tool in the former's hands, and external *power* assumes the place of the highest good—for the species, to be sure, not for the individual. This would be, then, since there is no limit to it, a "utopia" of permanent self-surpassing toward an infinite goal. Science, the life of theory, would be much better suited for the role of end-in-itself, but it can be this only for the small band of its devotees.

Is there *moral* profit in this intrinsic progressivism of science and technology? The question is germane to our subject, since we connect traditionally some ethical meaning with the idea of progress—nay, such a

meaning, as we have seen, predominates in the original, individual de-
notation of the concept. Thus we ask: Do science and technology through
their progress contribute to the general moral betterment? Surely, both
are not without a link to morality. Let's take *science* first. Seeing that
dedication to knowledge is an ethical good in itself, we may reasonably
assume that science—as generally a life of thought firmly committed to
truth—*can* have a moralizing, ennobling influence on its practitioners,
making them better persons, not only better scientists. Strangely and
regrettably, it does not always do so. But where science does act that
way, it does so not through its *progress* and, anyway, not through its
results at all, but through its actual performance in the scientist concerned—that is to say, through its abiding spirit; and in this, the later
scientist, however far ahead perhaps by grace of progress, has no edge
over the earlier. The ethics and psychology of the courtship of truth are
timelessly the same. Also, their inward gains do not affect the general
public.

That general public, however, that is, society at large, *is* affected by
everything which *technology* releases into the world, therefore indeed by
its progress, which is a progress of results. Now, about the complexity
of these results—as fruits for human consumption and as shapers of the
human condition—all that can be said is that some have a moralizing and
some a demoralizing effect and some possibly both, and I don't know
how to sum up a balance sheet here. Only the ambivalence itself is beyond
question. Should it be the case, however, that the long-term reshaping of
the conditions and habits of life will lead to an alteration of the human
type (not a groundless thought, considering the plasticity of man), then
this alteration will hardly go in the direction of ethical-utopian elevation.
The preponderant vulgarism of the technological blessings alone renders
this more than unlikely, even without considering the relentless disen-
franchisement of the individual through the technological order as such,
with its physical and psychological mass compulsion.

6. *Of the Morality of Social Institutions*

Whereas in the case of science and technology we could unequivocally
speak of progress, even of potentially endless progress—perhaps the only
extended sequences in whose nature it is that later states as such surpass
the earlier—the picture is much less clear in the field of sociopolitical
order, which has so much closer a relation to morality (and also, until
recently, provided much more of the material for historiography). Reflecting about the difference, one is tempted to posit it as a general rule
that, the more closely a phenomenon of collective life is related to morality, the less certain is "progress" in it as a natural concomitant of
change. That which is ethically more neutral and measured by purely
objective criteria, where every "more" means a "better," is plainly more

amenable to cumulative improvement; in short, "ability" more than "being." Yet, there *are* better and worse forms of political, economic, and social order; and apart from being in themselves more or less moral, that is, in agreement with moral norms, they also set better or worse *conditions* for the moral quality—the "virtue"—of their members.

a) Demoralizing effects of despotism   That these two things—goodness of the system and goodness of the individuals—coincide is not *per se* certain. As an effect of the system on individuals, it is rather more likely on the bad side than on the good. A despotic regime, for example, already as such conflicting with certain ethical values, also tends to corrupt in different ways the wielders as well as the victims of its arbitrary power (who may often happen to be the same persons). To stay pure in such conditions demands extraordinary strength of soul—more than one has reason to expect from the average person or has a right to ask of any person. We would not wish to do without the sight of the great witnesses to truth in the pageant of the past. In them it is revealed to what moral greatness "man" can rise, and but for them we should never know. Yet nobody will, for the sake of the precious flower of martyrdom, approve the soil in which it alone grows. On the contrary, its example bids us to strive for a state of things where its like is not necessary. (Plato searched for a polity in which a Socrates would not have to die.) Virtue need not be easy, but its cost must not be so high that it overtaxes the means of most. Also, one would wish its light to shine in itself rather than in contrast to the darkness of vice, although it shines brightest then—at its utmost, in an unearthly glory. But here, we speak not of the few but of the many, of the average condition of the countless members of the commonwealth (with whom, after all, utopia is mainly concerned). And there, considering how we humans are, the sober proposition holds that virtue needs to be encouraged and must not be discouraged. Least of all must vice be encouraged, and that happens, for example, in despotisms, and in totalitarian despotisms most totally: among the rulers, they encourage the vices of arbitrariness and cruelty; among the ruled, cowardice, hypocrisy, sycophancy, betrayal of friends, callousness, at least fatalistic indifference—in short, all the vices of fear and of survival at any price. And if not all of these are always vices (= the indulgence of active viciousness) but, under the circumstances, rather weaknesses of which the safe bystander dare not be the judge, it is in any case the ignoble and shameful sides of human nature that are here rewarded, and the better ones on which heavy penalties lie. Not that they are thereby made impossible (this is the ever-repeated, deeply moving lesson of the centuries, not least of our own), and the sacrifice of the upright is all the more impressive for it. It is also not in vain, for it saves our faith in man—

if it becomes known. But this involuntary gift of tyranny to a doubting mankind cannot serve the giver as an excuse.

b) *Demoralizing effects of economic exploitation*   What we have illustrated with despotism goes also for other institutional aspects of society: that they create moral *conditions* which can be beneficial or harmful to personal ethics, predominantly the latter. Our second instance is the *economic* order, with which we move closer to the argument of Marxist utopia, which is concerned with other things than "freedom." Taking the cue from the Marxist critique of capitalism, we can say that conditions of *exploitation* are immoral in themselves and also demoralizing in their effects; and, here again, for the winners as well as the losers. The exploiters become guilty simply by being this, and moreover, since nobody passes his working days with a feeling of guilt, they suffer on their souls the moral blemishes of insensitivity and a lying conscience, without which they could not play their roles successfully. For the rest, they may be blameless in their private lives and not without tender sympathy of sufficiently restricted scope. Also, entrepreneurship has always required and bred its own peculiar virtues; but in a morally wrong basic setting, those virtues too are stained by what they are employed to serve. Thus, the Socratic proposition that the wrongdoer first harms himself, because he makes his soul worse, is true also of the "exploiter," even when he is this simply by membership in his class and not by personal choice. To reduce here the individual factor in the causal net as much as the case requires, let us say that the *system,* through its overall distortion of roles, is morally harmful even to its beneficiaries, the exploiters themselves.

And the exploited, who are of course our greater concern? Are their souls also made worse through their being exploited? Does their moral potential suffer diminution by what is done to them? This is a question in whose consideration the freedom of the soul to rise above oppressive circumstances can be both underestimated and overestimated. Much depends on the degree, less perhaps of the injustice as such than of its objective consequences: beyond a certain measure, there is no question that these will destroy that inner freedom in their victims. Socrates taught that, for the agent himself, doing wrong is more harmful than suffering wrong. Well, so long as it is merely a matter of "wrong," done or suffered, thus a matter essentially existing in the perception of the parties, the proposition may stand. (Even there, taking psychology into account, we must perhaps believe Nietzsche more than Socrates.) But the *objectified* wrong creates a new, external causality, and we are inquiring about *its* moral harm to the suffering side. And we ask the question not, as Socrates did, for single actions committed and suffered here and there, but for the constant effects on the victims of a *system* of injustice. (We must know about this so that we can know about the effects to be expected from a *change* of the system.) And there the main point is that, in a system of

ruthless exploitation, those objective effects mean abject poverty with all
the degradation, external and internal, which this entails. We think of the
condition of the English working class in the early stage of the industrial
revolution, or the lot of the rural proletariat in some present agrarian
societies, let alone the misery of urban slums.[7] It needs no saying that
conditions like these curtail the whole human being—that poverty im-
poverishes further, not sparing the moral life. What force and fear do in
political despotism, material want and its tyrannical desires do here: where
they do not stifle the sense for "virtue," they make its price too high.
Bertold Brecht's words ring in our ears: "First comes the feed, and then
morality"; "But circumstances won't permit, alas."[8] We shall return to
the philosophy, implied here, of the power of "circumstances" and of
morality being conditional upon them, since it will be of more than negative importance for utopia. For now, we merely take note of the banal
truth that the conditions of a bad economic order, like those of a bad
political order, can hinder the individuals in achieving "goodness" of
whatever kind; and we need not dwell on the question of how much of
it the miracle of the soul may yet make possible, in spite of them. The
negative side of the evidence is clear.

c) *The "good state": political liberty and civic morality*   Is the positive
side equally clear? Namely, that the good economic and social order, the
"good state," begets also good people? That the removal of obstacles—
certainly a first priority—brings ethics along with it, so that with "feed"
indeed there comes "morality"? Is man good, if only you let him? (The
questions are not quite identical, since "causing" and "permitting" are
not the same.) Here we enter a less certain territory—the perennial fate
of ethics, in which the negative is so much clearer than the positive. I
hasten to add that no skepticism at which we may arrive concerning the
positive side, that is, the advent of the good, relieves us of the duty to
abolish the cause of the negative side and replace the bad conditions with
better ones to the best of our ability. That which is a moral scandal must
be done away with, even if we do not know for sure what in the end we
shall get in its place. To this first duty, therefore, neither doubt nor faith
in man's moral nature makes a difference. But it does make a difference
for Utopia and generally for the grand designs which set their sights
beyond the correction of ills. And so, the ancient question about the
relation between the good state and good citizens shall be posed here
once more.

Here one recalls the extreme saying of Kant that the best state is the
one which, even composed of devils, would function well—and Kant
(unlike Hobbes) meant: function according to laws of liberty![9] It is the
same state that would be best also for angels; that is to say, it is itself
ethically neutral. This stands in conscious opposition to the political the-

ory (in part also practice) of classical antiquity, which held that the good polity is to be a nursery for the *virtue* of its citizens, on which conversely its own well-being depends. The state can only be as good as its citizens, and more was meant by their goodness than keeping within the law. (Hence, for example, the official supervision of morals in republican Rome, extending right into the households.) Something of this moral view of the state survives in some modern thinkers (Hegel), and occasionally even, in moments of Jacobin-republican fervor, it is conjured up with the battle cry of the "citizen virtues" which must adorn political liberty. But in general, since Machiavelli, this classical idea of the state as a "moral institution" has progressively vanished from modern political thought. Dominant in the Western world became the liberal view of the state as a utilitarian institution which should protect the safety of the individuals, but within its bounds leave the widest room to the free play of forces and, especially, interfere as little as possible in citizens' private lives. The concept of rights to be assured overshadows that of duties to be demanded. What is not forbidden is allowed, and fulfillment of the laws consists in not transgressing them: reacting to transgression, the power of the state turns active. What otherwise the individual does with the publicly protected free space of his existence is his own business and none of the state's. The best state, accordingly, would be that which is noticed least (the "night-watchman state"). Compared with this liberal conception, the communist goal for society—the pervasion of every individual life by the public interest, its adaptation to it and conscription for it (its best example in miniature perhaps being the Israeli kibbutz)—looks almost like a return to the ideal state of antiquity.

We need not choose between these abstractions. For our concern here is not what public systems think of themselves or how they are thought of, but what effects they produce—in particular: how they affect the moral being of their constituents to the good or to the bad. Now, from what we have said of the corrupting effects of despotic regimes, this much at any rate follows: that libertarian regimes do at least avoid *these* causes for corruption. Beyond this truism, they are for their part, of course, full of problems, and the principal problem is the confounded liberty itself which, self-evidently, is by no means a freedom only for the good. Each enhancement of freedom is a great wager on the odds for its good use over against its bad use, and only he will be certain of these odds who is convinced of the natural goodness of man (not to speak of the distribution of *insight* even with the presence of good will). Yet, even he who is not so convinced must dare the bet of freedom, for freedom is an ethical value in itself and worth the risk of a high price. How high a price and how great a risk? There is no *a priori* answer to this question and no recipe for determining it. The sense of responsibility and what wisdom one has must answer it according to the circumstances. Most often it will be wisdom after the fact that must do the judging, for only the wager first accepted reveals the sequel in the try. In any case, a libertarian system *permits* the virtues that can flourish under liberty only, and possessing them rates in general more highly than avoiding the vices also made possible by liberty and the avoidance of which a system of unfreedom can set against the claims of liberty. Relevant in this balancing is the consideration that taking the risk of independence in following one's own judgment is already a virtue in itself and more becoming to man than safety in the shelter of prescription. Thus (surely with the Western bias on our side) we may assert that in all areas of human activity a system of freedom, as long as it can guard itself against its own excesses, is for ethical reasons to be preferred to an unfree one—regardless of whether the latter can serve certain and even important communal interests more securely or efficiently. The same holds for other alternatives. A government of law is better than one by arbitrary decree, equality before the law better than inequality, right of merit better than right of birth, open access to merit better than one presorted by privilege, control of one's own affairs and a voice in public affairs better than their permanent surrender to the trusteeship of magistrates, individual variety better than collective homogeneity, therefore tolerance for otherness better than pressure for conformity—and so on, and so on. What is each time called "better" in this somewhat banal enumeration *may* also be better technically, that is, more effective in discharging the social tasks that a public order must look after. But this is not necessarily so. Our point was that they are ethically superior to their opposites, which may surpass them in efficiency here and there, or even on the whole.

No doubt, then, without defining "the good" itself, there *are* better and worse social systems; and insofar as the better ones, like government of law, broadening of civil liberties, etc., are the fruit of developments and often of protracted efforts, and so, in plain fact, appear *later* in history than the others, we can here, surely, also speak of *progress*. But we must add immediately that their being better, as measured here, by no means also guarantees their durability. To the contrary, precisely *in* their superior properties, most of all that of freedom, they harbor the germ of contradictions, internal crises, possible degeneration, and even total somersault into their opposite. ("Retrogression" would hardly be the right word for the last case, since the new tyranny would bear very novel, "progressive" traits.) In one word, the ethically better are at the same time the *precarious* systems, and for this reason alone they are unsuitable for Utopia, whose first *formal* requisite must surely be that it can be trusted to endure. Demonstrably at any rate, if we recall the horrible "somersaults" at the height of history, the later is here *not* necessarily the better, even if profiting precisely from technological progress in the means of power and thereby perhaps proving superior to the displaced system.

d) The compromise nature of libertarian systems We must go a step further in qualifying the perfections of liberty. Durability apart, not even in their *content* do these more desirable systems represent the "ideal" of pure, unmixed desirability. This is quickly seen when we prolong the foregoing list of "*A* is better than *B*" by the following, no less persuasive list: security, public and private, is better than insecurity, therefore firmness of the established order better than softness; enforcement of its laws (provided they are not bad laws) better than inviting ease of circumvention; effective police and judiciary better than ineffective ones, impeded, for example, by too much consideration of individual rights and liberties. And beyond this tedious complex of "law and order," but still connected with the principle of "security": legally ensured provision of the basic necessities of life for everybody is better than permitting want and deprivation through the caprices of the market. Therefore, public control over the distribution of the social product (including immaterial products like health and education services and even workplaces)—a control inevitably extending to the sphere of production and its overall planning—is better than a distribution left to the mercy of individually possessed or lacking purchasing power, and than a production left to the fiat of untrammeled competition; therefore also the "welfare state" overseeing the economy, better than the individualistic "sink or swim" system of free enterprise, and so on—and extending through all this, *including* the goods of the first list: *stability is better than instability*. Now, one immediately recognizes that, of these two, equally evident "better than . . . " series ("freedom better than unfreedom" on the one hand, "stability better than instability" on the other hand), not everything can be had *together* in the same degree, that rather certain goods of one side can be had only at the expense of certain goods of the other side—that here, therefore, *balance and compromise* between that which in the bipolar extreme is incompatible is the best one can realistically hope for. As in everything real, the law of compossibility is operative in this province too. Especially the *libertarian* systems, of necessity balancing between the internal threat of anomy and the external threat of remedy by (egalitarian or inegalitarian) coercion, are *in their very nature* built on compromise; and since compromise again by *its* nature is something imperfect and moreover, because of the premise of liberty, is fluid even in its imperfect reconciliations—that is, always forced to readapt itself—"stability" is once for all *not* the forte of libertarian systems. But Utopia is temperamentally the last to be patient with compromise and internal imperfection, that is, with half-measures, let alone instability. And since on principle only the nonlibertarian alternative can be had without compromise, whole and (at least conceivably) free of change, it follows that every *realistic* utopianism must put its money on *this* alternative. In other words, it must opt against the individual and for the collective (and so on)—that is, for something in other respects quite imperfect. That the barter of individual liberty does not appear to the utopian too high a price for what is gained thereby, nay, that he regards the sacrificed good as a mere illusion (a "bourgeois prejudice"), is understood, but does not bind our judgment.

7. Of the Kinds of Utopia

a) The ideal state and the best possible state   Having invoked the principle of compossibility, we see that we must distinguish between two quite different concepts of the best or "ideal" state: the best in the idea, in and by itself, irrespective of possibilities of realization—simply best and most desirable according to a wish-image of human happiness and freely imagined by this standard alone; and the best state *possible* under real conditions, a construct taking into account the limits of nature and the imperfections of men, who are not angels, but not devils either. Of both concepts, imaginary models can be drafted for contemplation, "utopias" as a literary genre—the ones radiant in their unblemished, ideal positivity, the others alloyed, perhaps melancholically so, with the admixture of the imperfections of nature and man. "Thoughts dwell with thoughts in easy company, / But hard in space do things with things collide" (Friedrich Schiller): true, but also the laws of space can be admitted into thought— hence the other sort of Utopia. The first sort is "U-topia" in the literal sense (Nowhere) and belongs in the Cloud-Cuckoo-Land of idle fancy; the other, whose first great example is Plato's *Republic*, is still Utopia in the sense that, although its *being* real would be possible, its *becoming* real in the confused flux of human affairs requires such a coincidence of lucky circumstances that its coming about is not to be reckoned with. (Better are the chances of its lasting, should the improbable have occurred after all, but even they are uncertain in the long run.) But by itself, the model is meant to be realistic, that is, capable of existing in the world as it is; and since it includes the limits of man in the data of its calculus, and on the other hand, for the lucky achievement to endure, it cannot afford much in the wager of fickle freedom, it is natural for this class of "realistic" utopia to have a strong authoritarian and paternalistic component in its makeup (for which Plato had to take assorted beatings to this day). "Utopia" in the literal sense is this kind too in that it is not meant as a plan for political action, except in the unlikely event of the "lucky coincidence," about which itself next to nothing can be done. But as a basic meditation on what is politically possible and worth aiming at, this kind of utopia is not idle as the other is, and can even serve as a compass in the labyrinth of political practice.

b) The novum of Marxist utopia   The dichotomy of utopias I have just outlined fits the evidence of an earlier time. Our modern utopians fall in neither class. They intend to *bring about* their utopia in all earnest, holding that history has made it now a realistic hope, promotable by human action, which it never could be before. Thus, theirs being not a dream in the clouds, one should expect them to offer a "utopia" of the second, resigned kind: an exercise in realism by people who want to be realists more than anything, and nothing less than to be judged "idealists." But this is by no means their option. A novel third choice beyond the pre-Marxist alternatives is the particular trademark of the revolutionary utopianism of the present, and this we must now introduce into our discussion of progress, in anticipation of our next and last subject.

The Marxist can answer our previous reflections with two connected rejoinders: (1) *revolution* is ignored in them; and therefore (2) no account is taken of the radical *change* which it will effect in the human equation, whereby the analogy of the past no longer holds for the future. For in our case (so he can say, first), the *coming* of the utopian state of things—of socialized production in a classless society—is by no means entrusted either to lucky chance or to the immanent trend of progress, much as past progress has *prepared* that coming dialectically and at last made it *possible* realistically: rather, for the decisive turn, there exists the way of revolution as a conscious and violent intervention in the course of events—which is tantamount to human *craft* on the grand scale taking things in hand and forcibly fashioning the new condition out of them. Accordingly, this new and consummating condition, though the one most fitting for "Man" writ large and in this ideal sense most "natural" to him, will be a creation of art superseding nature. And this creation, second, will be so novel in the *conditions* of human existence, and so liberating for the hitherto inhibited human potential, that no comparison with the past still applies and all former history pales to prehistory. The true history of mankind, nay, "true man" himself, only begins with this juncture. How this "true" being will concretely look—the *content* of the vita nova due to reveal itself after the rebirth—of that, no description can be ventured from the state of untrue being in which we are still submerged. But two things can be said already: negatively, that the anti-utopian objections drawn from the "nature" of man as known hitherto are "then" no longer valid, because that "nature" was itself the product of the circumstances now superseded; and positively, those circumstances having been inhibiting and distorting, the state free of them will at last also free man for the uncurtailed being of himself. The truly human history which will only then begin is that of the kingdom of freedom existing for the first time. To its content, not describable in advance, we revolutionaries look forward with faith in the new man.

Thus speaks the new millennialism of the postreligious age. We shall critically inspect it in the next chapter.

6
A Critique of Utopia and the Ethic of Responsibility

The closing sentences of the last chapter provide the theme for this one. A critique of utopia has become necessary with the seeming possibility of its realization. For the first time in the annals of man, thanks to the powers of technology, the dream appears to be capable of turning into a task, and Marxism has seized on this novel chance to give its political gospel eschatological exaltation and pragmatic credibility at the same time. Nothing could tempt the might of Prometheus unbound more than the dream of the highest earthly good believed within its reach, and nothing can become more dangerous to mankind than a mistaken pursuit of it. In Marxist utopianism, technology is put to its most ambitious test. A critique of the one is thus a critique of the other in surrealist magnification.

The spirit of Marx's technologically underpinned utopia is thoroughly humanistic, and this—to give Marxism its due—is the true idealistic meaning of its materialism. What hitherto counted as the nature of man, so the doctrine goes, was the product of obstructing and distorting conditions. Only those of the classless society will bring to light his true nature; and with its "realm of freedom" only will prehistory end and genuine human history begin. This is strong stuff. Religious faith alone had hitherto held out similar hopes: a consummating transformation of man, even of nature, with the coming of the Messiah or the "second coming" of Christ; a second creation perfecting the first, free of sin; the new "Adam" raised from the fall of the old and immune against its recurrence; the *imago Dei* on earth at last shining forth in its intended purity. The atheist analogue has its obvious differences from this model. A secularized eschatology of the new Adam must substitute worldly causes for the divine feat that was to work the miracle of transformation in the religious creeds. The secular equivalent of that work is now assigned to the external *conditions* of life, which can be politically instituted, namely, through the socialization of production and distribution in a classless society. To bring this about is the task of the revolution, which here assumes the task of the divine intervention, and all the rest can be left to the effects of the resulting new order. With no outpouring of the Holy Ghost, by its own molding force, will the right order of things make the pentecostal miracle happen. Everything is therefore focused on the revolution and its stages, that is, on the process of *bringing about* the new order. Contrary to the earlier utopias, it is the *coming,* not the *being,* of utopia about which Marxism has something to say. Its "being" it can no more describe in advance than religious eschatology could the Kingdom of God—except, negatively, that as sin in the latter, so the evils of class society will have *disappeared* in the former. Rightly, neither draws an imaginative picture of the positive manifestations of the new state of things[1]—which was the very matter for pre-Marxist utopias to dwell on, while *they* left the "coming" veiled in darkness. This is indeed a novum of Marxist utopianism and one of the features which mark it as secularized eschatology and an heir of religion. (Another feature is the doctrine of the "sinfulness" or radical provisionality of all previous history.) But what a demand on faith! Belief in God, once there, may well entitle the believer to affirm in advance ("sight unseen," i.e., even without an *idea* of what it consists of) a future transfiguration of man wrought by him as certainly the best state possible, and hence also to accept the "labor-pains of the Messiah," the convulsions of the Last Days, as its price. In the totally unknown, faith alone reigns, and one cannot remonstrate with it. But where the "labor-pains" are the work of man, namely, world revolution, and likewise its result, the classless society, and this is to consist of the familiar stuff of our world, there the belief in its "saving" power, conceived of as a thoroughly this-worldly causality, must submit to this-worldly examination. The "sight unseen" of the transcendental miracle cannot be invoked on behalf of *its* promised hereafter, especially as we ourselves are to initiate it. But "seen" and inspected, it also poses the question of price which cannot be asked there.

The "faith" that needs to be (and can be) examined here is composite: the faith in the "power of circumstances" generally and in man's being wholly their product; then the belief that there can be nonambivalent circumstances, that is, such that are *only* to the good; then that, placed in them, man will be as good as they are, because they permit it to him; and finally, that this good man has never yet been, because past circumstances did *not* permit it—that hidden "true man" is still to be revealed. The last point is crucial for Marxist utopianism, because its passion lies not in the intent merely to improve conditions that cry out for it, not in reform, but in the promise of a heightening *transformation of man* through conditions never known. This has decisive influence on the question of what steps may be dared, what risks taken, and what price exacted for such a transcendent prospect. Here the spell of Utopia can well become an added danger, precipitating what should be held in check. For when it really is a case of bringing inchoate "man" first to himself, then no price may seem too high and no ruthlessness forbidden.

I. The Wretched of the Earth and World Revolution

It must be interjected here (and this will detain us a while longer from examining the utopian idea itself) that no such sublimity of the revolutionary goal was needed for moving the masses to whom the call was originally addressed: the industrial proletariat, pauperized and (according to the theory) doomed to ever-greater pauperization under a capitalism growing ever richer at its expense. For the "wretched of the earth," who have "nothing to lose but their chains," it required no dream of the new man, or of any kingdom of heaven on earth, to make them wish to obtain *deliverance* from their intolerable condition through a redistribution and socialization which they perceived to have become possible now and enforceable by their solidarity. Deliverance as such is the dream of suffering, and any approximate equalizing with their former exploiters, a fair share in the surplus they themselves create, could appear to the sufferers as the boldest "utopia": surely as sufficient for the attempt at revolution. Its failure could make things no worse, and its success only better—no matter whether ultimately and in all things "good." To this powerful motor of need, the urge of suffering for deliverance from itself, the prophets of socialist revolution could appeal, while *their* goal went far beyond this. And that motive, we must add, fully suffices by every human norm, moral and pragmatic, to justify violent revolution, if there is no other way to its satisfaction.

1. Change of the "Class Struggle" Situation by the New Planetary Distribution of Suffering

That still holds today for the "wretched of the earth." But they are no longer located, as a group and mass phenomenon, within the advanced industrial nations (in the capitalist "West" even less, if possible, than in the communist "East"), but outside them in the underdeveloped, ex-colonial countries of the so-called Third World. And there they exist as entire, impoverished peoples, not as oppressed classes in economically rising societies.[2] That changes the addressee and meaning of the preached revolution, which here becomes "world revolution" in an entirely novel sense as essentially *an issue of foreign affairs*.

a) Pacification of the Western "Industrial Proletariat"
In the successful, industrial nations themselves, which harvest the fruits of the technologically heightened production, the situation of a worker proletariat defenselessly thrown to the mercies of the market is noto-

riously long since a matter of the past. Where powerful labor unions and a management dependent on bilateral contract face each other across the negotiating table, one can no longer speak of unilateral exploitation, even if the finally arrived at benefits remain unequal. A power contest regulated by statute, with a roughly balanced division of forces, decides the workers' share in the proceeds of the business and therewith the eventual apportionment of the social product as a whole. If negotiation fails, then the (still peaceful) strike holds to ransom not only the immediate opponent but broad sectors of the economy as well and, in the case of vital services, even the entire public. Often it is "capital" which must yield. The economic gain to the labor class from this "reformist" development, which was wrested from the other side in part by force, in part by the latter's own prudence (and even morals), and politically secured in law, is well known. The living standard of the worker in the capitalist West of today surpasses that of most of the modest burghers and peasants of the past prior to the onset of proletarization;[3] and present conditions would have looked like paradise to the victims of pitiless, early industrialism. I doubt that they would have expected more from a successful revolution. Moreover, "socialist" aspects of public welfare, such as health and old-age insurance, have eliminated a good part of the insecurity of existence of earlier times. *Inside* this world, the previously cited extreme of "morality" being suspended for lack of "feed" has disappeared as a group destiny.[4] These fruits of a mostly peaceful process have indeed made the working classes of the West (now having a vested interest in a system whose functional mechanics embodies inbuilt means of pressure for their side) largely unsuited as candidates for the revolution. For want of an oppressed class, the revolution is called off. Its true protagonists, who care little about mere "improvements" in the condition of their adoptive class, but rather are suspicious of them as a sort of "bribery," have therefore with good reason always regarded "revisionism" (= reformism) as their worst enemy. The true "utopia" animating them becomes the more important for being now the cause of a splinter elite of radical idealists from the upper classes which, ironically, finds inside their society as a whole the least response among the intended objects of their mission. Their natural tactical allies are nowadays altogether elsewhere—among the "wretched of the earth" in other parts of the world. Naturally, here too it is not premature pacification through improvements in their condition, but mobilization of their unassuaged revolutionary potential, at which the true utopianism (rendered homeless in its own land) must aim. But the actual and viable pacification, on the whole attained in their prosperous native lands—which in some sense is, after all, a (preeschatological) victory for their cause, insofar as it would hardly have come about without its menace (and certainly not without some persuasive force of its moral eloquence)—shows that the nonutopian, and eminently rational,

elements of the original conception can also be achieved by nonrevolu-
tionary, gradual transformations of the ruling system which in the net total
tend in a "socialist" direction.[5] Also in other areas, as has often been
noted, the internal transformations of capitalism lead to structural con-
vergences with the existing "communisms" born of revolution (but still
pre-utopian). In any case, increasing socialization in the form of the modern welfare state seems presently (and foreseeably) to be the general
tendency of Western societies, in a fluid balance with the principles of
liberty (and the clamorings of unreason). All this is a far cry from perfection, but of this one only speaks in utopia.

b) Class Struggle as a Struggle between Nations

But there is no lack of the "wretched of the earth," who are as necessary
for the Marxist revolution as water is for the mill-wheel (or fuel for the
explosion). As we said, these are the truly pauperized populations of the
"underdeveloped" parts of the world, within which, indeed, once again
oppressed classes exist. But the poverty of the whole is so great and so
endogenous that even a sweeping away of the thin, local, parasitical up-
percrust would alter little. Those geographic entities as a whole are today's
"oppressed class" in the global hierarchy of power and wealth, and their
"class struggle" must necessarily be fought out on the international plane.
Their motor force, impelled by bitter want, thus could well be harnessed,
beyond their own nearer goals, for the utopia intended in the world revolution. And here, too, all the instincts of philanthropy and fairness in
the privileged world itself, which likewise in themselves have little to do
with utopian aspirations, could be mobilized in the latters' aid. But the
situation is basically different from that of internal class struggle on the
national stage, that is, inside one and the same functionally and territorially coherent society. Everything here is much more mediate and capable
of insulation. Exploitative guilt of the privileged nations is only partially
involved ("economic imperialism"). Unpropitiousness of nature has its
heavy hand in the game, and even indigenous, historical-anthropological
disposition may be a factor in staying aloof from civilizational progress
(where such disposition is not, circular as such relationships are, conversely a result of that staying aloof). Insofar as *ethics* is supposed to
motivate the aid from without, it will have to be a more voluntary, mag-
nanimous, and at the same time softer kind than an ethics under the
conjunction of guilt, justice, and neighborliness in one's own house. Regarding the direct appeal of misery to the sensibility, it is a fact that
distance screens the latter sufficiently to make it let pass the known
starving of remote populations which would give it no rest close by. The
entirely valid "Charity begins at home" easily leads to its ending there;
and direct responsibility has for the sentiment its limits in nearness. So
it is in terms of individual psychology. For the group, the political col-

lective, of which from the start it cannot be expected that it will be "noble, helpful and good," but which must here be the real agent, enlightened *self-interest* takes the place of personal ethics, and that interest indeed commands not only a palliative mitigation of alien indigence by handouts from one's surplus, but even lasting sacrifice of self-gratification for the sake of eradicating world poverty from its roots. What is missing in the "wealthy" nations is therefore not so much selfless kindness as the enlightenment, that is, *farsightedness,* of that very selfishness whose inclination to nearsightedness is so indelible, since the "self" is always represented only by those just living.

Farsighted self-interest would here be twofold: the (in the long run) better return effect of a healthy world economy upon one's own, and the fear of pent-up need exploding into international violence. The latter can assume the traditional form of wars between states (e.g., a coalition of "insurgent," impoverished peoples, possibly under the direction of, or at least armed by, a third power), or more likely the new form of international terrorism (without determinable national responsibility) to extort from the lands of surplus economic tributes to the lands of want. In any case, the "class war" would inevitably degenerate into national war of the old style, if not into race war, and call forth on the attacked side (including its working class!) all those impulses of national solidarity which would smother whatever sympathies with the other side were previously alive in its midst. In the end and at its most deadly earnest, the appeal to force would probably go against the weaker side—if perhaps with the aftermath of a belated help by the victor to the vanquished. No real predictions are possible here. But the whole prospect of international anarchy opening up here is frightening enough to make a wise policy of *constructive* prevention appear as the best for one's own long-term interest; and the same we are told by the more peaceful, purely economic consideration.

2. Political Answers to the New Terms of Class Struggle

a) *Globally Constructive Politics in the National Self-Interest*

Such a constructive policy, however, is faced with the greatest problems of which we skip the specifically political and preliminary one: how to obtain domestic assent to that policy on the giving side, so long as a condition of voluntariness exists. What, taken by itself, would be constructive? What comes most readily to mind is a replay of the industrial revolution among all backward peoples, avoiding the social sins of the original and primed with investment and knowledge aid from the outside: thus the *addition* of comparably productive capacities everywhere to those already existing in the "seed countries"—thus the spreading over the globe of the highly intensive industrial technology hitherto concentrated there. We indicated earlier that the globe, already showing signs of strain, could probably not endure such a multiplied attack. Where the limit lies, only time can tell;

but one should not let it come to that. The alternative would be a partial relocation of capacities from the "high-pressure" to the "low-pressure" regions, so that in sum the global strain on the environment stays within bounds. Such an assimilation of levels, whose purpose is the raising of the lowest, naturally means a lowering of the highest: the cutting back of production capacities there with a corresponding shrinkage of consumption capacities owed to them: and there, the preliminary, political problem becomes acute indeed! To be sure, objectively there is no doubt that in the developed countries (overdeveloped by reasonable norms of consumption) there is comfortable room for a trimming which would leave us still with far more than our grandparents and even our parents had. But the subjective reaction to such an imposition without a visibly present necessity thereto is another story; in America, for example, spontaneous resistance (again including the working class) would be virtually certain. Despite this, I believe that the solution lies in this direction—freely if possible, compelled if necessary. It is clear, however—and this is our principal problem—that *every* constructive solution requires a massive infusion of technology (the sheer numbers of the earth's present population excludes a return to earlier conditions), and the wounds *thereby* inflicted on the environment demand further technical progress for their healing, that is, more and better technology already from sheer defensive necessity. Aggressively it will at the same time aim at *pushing further back* the aforementioned limits of tolerance of the natural environment: this, too, double-edged again in each success, ever more precarious, and naturally not endless. The here reigning dialectics of a progress which, in providing solutions for the problems it has created, must create new ones and thus becomes its own compulsion, is a core problem of the ethic of responsibility for the future of which we are in quest. That problem will at some point force the idea of progress to a retrenchment from expansionist to "homeostatic" goals in man's relation to the environment (which will still set sufficient tasks to technology as such for unceasing further development). At the moment all we can say is that in the danger zone we have entered with our technology and in which we must henceforth move, *caution* rather than exuberance is needed, and the spell of utopia—our present theme—is the last thing which must be allowed to dim the clarity of vision which is here required. Admittedly, this sentence is predicated on our not believing utopia to be possible.

b) The Appeal to Force in the Name of Utopia

But for the champions of utopia, things look different, and no scruples against "moving the Acheron" of the Third World to world-revolutionary violence need detain them. For, first (as we put it before), when it is really a case of bringing inchoate "man" first to himself, then the antecedent costs do on principle not matter, and the most colossal mass extinctions can appear as a necessary, alas painful, but beneficent surgical operation, if the final *regnum humanum* on earth cannot be brought about otherwise. (What now exists is on its merits—by hypothesis—expendable.) Second, utopian ideology contends that the present perils and limits of technology will *then* no longer apply: not only because technology will then, freed from the irrationalities of the profit economy, be employed more wisely, but also because, freed from the social inhibitions of its still unplumbed potential, it will only then, as the finally unbound Prometheus, attain to the highest possibilities of itself. And these have no other limits than those forced on it by an imperfect society—neither limits to its own, always self-surpassing, creative *abilities,* nor limits of *nature* to their ever-expanding application: the potential of nature, too, is *in itself unlimited* and only waits for utopia with its more ingenious art of invention to unlock it fully. Moreover, third, even if success is not regarded as certain, yet the real chance for it (which exists if only the goal is *possible* in itself) joined to the radical condemnation of the existing state as unworthy of man, lifts the prohibition we stipulated on playing banco with the destiny of humankind: with so absolutely obligating a goal—no less than *making true man possible* for the first time—one might feel justified to put at stake the whole which is anyway not worth the candle without it. If this is the either/or, it could seem right to dare the utmost in technological audacity, to earn either supreme bliss or supreme disaster.

Regarding the first point, we would only say, disregarding morality (questions of the humanity and inhumanity of the "price"), that those who have emblazoned on their banners the cause of the victims of power should be the last to appeal to the arbitration of power. Not only will the decision most probably go against them and they will then have forfeited the appeal to reason and other forces of reform, but also a decision for them could only lead to another system of power with all the blemishes of it. This is not to mention the fact that the devastation of the globe and the barbarization of man, which such an Armageddon would first bring in its wake, would for long delay the arrival of utopia just in default of the suitable (environmental and human) "stuff" for it. I repeat, *despair* may little care about this, but those who freely decide to make use of it must keep it in mind. Concerning the third point, the readiness for the "all or nothing," we say only that it would be *the* ethical, indeed metaphysical, deadly sin possible which mankind (as distinct from the individual) can commit, and that not even the highest *intrinsic* credibility of the goal could lessen the nefariousness of a risk of that immensity. That a self-styled *avant-garde,* in the arrogant certainty of their subjective dream, should impose it upon humanity is a totally intolerable thought.

But ultimately it is the intrinsic credibility of the goal conception itself which is philosophically in question, and a part thereof is already the credibility of the *material,* technological and ecological conditions for it, whose realization the second point asserted to be tolerated by the nature of things. In turning to an examination of this point, we leave behind the preliminary question concerning the human driving forces which it would be possible and permissible to mobilize for the revolution, and seek a judgment on the goal itself. Obviously already a negative verdict from the "material" (matters of fact) scrutiny of the external prerequisites would by inference settle the other two points as well; more so, of course (because by more than external fiat), would a negative verdict from the "ideal" (essentialist) scrutiny of the human condition substantively expected from those material premises in moral and psychological terms. In these two steps, moving from without to within, we propose now to undertake the critique of the utopian ideal by itself. The first step, not yet into the heart of the matter, falls in the cognitive field of things physical and is thus an exercise in natural science; the second, concerning the life that is to inhabit the house of utopically transformed nature—the step into the critique of essence—falls in the cognitive field of soul and mind and is thus an exercise in anthropology and philosophy. As is all too well known, it is the fate of philosophical knowledge that, although concerning what by its nature is ultimate, it forever remains open to the clash of opinions; whereas natural knowledge can achieve a large measure of objective certainty: *its* results then may well with one stroke also predecide the philosophical question from the angle of external attainability, namely, in the negative, over the head, as it were, of the intrinsic desirabilities which philosophy would have to examine. Nevertheless, the consideration of what on principle we can expect of and for man—of that too wherein, with all of his futurity's openness, the perpetual and unsurpassable *present of the humanum* consists—retains its irreplaceable value for the proper, namely, *free,* farewell to the utopian ideal in case the denial of the material conditions compels it: it could be free, of one's own will, even if, with a less than unequivocal verdict on external factors, it would have to be chosen nonetheless—in the case namely that the ideal itself were found to be a false god, a wrong object of hope.

II. Critique of Marxist Utopianism

A. First Step: The Physical Conditions, or: Of the Possibility of Utopia

1. *"Reconstruction of the Planet Earth" Through Untrammeled Technology*

The first requirement for utopia is material *plenty* for satisfying the needs of all; the second, *ease* of appropriating the plenty. For the formal essence of utopia, as we shall see, is *leisure,* and leisure can only exist with comfort, with an assured supply of the goods of life and a minimum of toil in obtaining them, since leisure is formally just *freedom* from toil in the service of need (or of wish fulfillment in general). A perfected technology can procure both of these—abundance and its toil-free command—in magnification of what in many places it has already begun: the first through the "reconstruction of nature" or of "the planet Earth" which causes it to lavish its hitherto niggardly doled-out treasures, or supplements its insufficiencies artificially; the second through the mechanization and automation of the labor processes which in the past had consumed human strength and time. Both coincide to some extent. For the "liberation" of the hitherto grudgingly granted natural treasures can *eo ipso* be only attained with enormous aid from machinery, hence at the same time with a liberation from human toil. This indeed is in full swing already in the unredeemed world, and we need not pursue the completely hypothetical question of whether only a Marxist society, as its heralds contend, can bring about the greater technological miracles necessary for the still greater, and finally total, utilization of nature and unburdening of man. The past course and present state of science and technology, wherever they thrive indigenously, promise of themselves further progress as virtually certain, and occasional lucky "breakthroughs" as quite probable. And although one must not bank upon the latter, surely not on specific ones, yet one can in general assume with good reason, however the particular technological society might look in other respects, that there will be a continual increase of capacity and application in all possible directions and for an indefinite future. The scientific-technological *possibility* for this exists and on the whole also the tendency in this direction. The advantage a Marxist society *might* have here over a capitalist one would hardly lie in superior science and technological inventiveness (the record to date tells rather the opposite), but in these two respects: a socially better selection and steering of the *directions* of technical progress (e.g., avoidance of much wasteful nonsense of consumer titillation), thus a better use of productive capacity; and, above all, a socially better, juster *distribution* of its fruits. The latter (especially with an assist from the former) could already with the yield of present technological economy greatly mitigate today's deprivations on our earth, and there is no doubt that the problem is in part not technologic-physical but economic-political. But only in part, and at best for the modest goal of securing a *tolerable* subsistence for the world's population at its present size. For more ambitious if still far from utopian goals, nay, even for holding the line on a pitiful status quo with a rising world population, *increased* global production and heightened, more aggressive technology must be the order of the day—and *a fortiori* for the universal "leisure-*cum*-plenty" economy envisioned by utopia: an altogether enormous enhancement by several orders of magnitude of both technology and its onslaught on resources. Let us, for argument's sake, grant the purely technical capability with all the required surpassings of present technology expected from the future—although the success-induced and by now habitual *confidence* in this, especially in a permanent exemption from the law of diminishing returns, is exaggerated. But then, precisely on the most *optimistic* view of the continuing rise in human *power,* the real question only begins.

2. *The Limits of Nature's Tolerance: Utopia and Physics*

The question is how "nature" will react to this intensified onslaught—nature to which it makes no difference whether the assault comes from "right" or "left," from the bourgeois-liberal or the Marxist camp: as surely as the laws of nature are not a bourgeois prejudice (though Marxist ideologues are sometimes inclined to that belief, and Stalin even acted on it in the matter of the laws of genetics). Thus one question posed by the utopian ideal is that of the limits of the tolerance of nature, in brief, *utopia and physics.*

The utopically postulated, double inexhaustibility—of human technology and of a nature responding to it—appears, for example, in these words of Ernst Bloch, the foremost prophet of Marxian utopianism in our century:

> Artificial fertilizers, artificial irradiation, are, or could be, on their way to incite the soil to thousandfold yield, in a hubris and unprecedented "anti-Demeter movement" whose synthetic boundary concept is "the cornfield growing on the palm of a hand." In short, technology per se is meant, and already not far from being able, to emancipate us from nature's slow and regionally confined labor on its raw materials. . . . A new raising of given nature to a higher state (a supernaturation) is due.[6]

In the last resort, the question to be asked here is not how much *man* is still able to do—there one may be sanguine for the Promethean potential—but how much of it *nature* can stand. That there *are* limits of tolerance nobody now doubts, and in our context the question is only whether "utopia" lies inside or outside of them; and that depends on its own numerical dimensions—brutally put: on the size of its membership. As to the limits of tolerance themselves, they become first noticeable to man when the detrimental "side effects" of his interventions begin to dim their benefits and then threaten to overtake them. And the limits are crossed, perhaps beyond the point of return, when the one-sided overstraining has set the whole system of countless and delicate balances adrift toward catastrophe in respect of human ends. (For nature itself knows no catastrophes.) It is a fairly recent and upsetting discovery that something of the kind is not only possible by the laws of physics but also, for the sharply defined spaceship Earth, lies definitely within the causal range of what man can do to it and in a considerable measure is already doing. This dawning truth of ecology puts a hitherto unknown damper on progressivist faith, socialist no less than capitalist. It is, remember, the nature of things around us which we consider here, not yet the nature of man himself, which will be considered in the second step. The more rigid limits of the former may fittingly come first in a critique of utopianism. What are they, and where do they lie? How far away as yet, or how close at hand already? Much of our present perception of this is necessarily guesswork.

The question as a whole lies in the domain of the infant science of ecology, and as to particulars in the fields of the biologist, the agronomist, the chemist, the geologist, the meteorologist, and so forth. In addition, it also ropes in the economist and engineer, the city planner and transportation specialist, and so forth. Only the interdisciplinary pooling and integration of all these will lead to the global environmental science that is needed. Here the philosopher has nothing to say, only to listen. Unfortunately he cannot borrow firm results for his purposes from the present state of the art. All quantitative predictions and extrapolations, even in the single disciplines, are at this time still uncertain, let alone their integration into the ecological whole, if such should ever be computationally feasible at all. Still, for various lines of progression one can indicate the kind of limits that lie in the nature of the case, and this can be instructive for judging at least utopian prospects, which everywhere bring the extreme into play. In view of the broad public discussion underway on these issues, a sketchy reminder of commonly known aspects will do.

a) Food and mineral resources   The three items around which extrapolations of needs and limits revolve are food, raw materials, and energy—and spread over all of them is the issue of environmental pollution. Bloch, with his "reconstruction of the planet," was mainly thinking of food production. Alas, his vaunted artificial fertilizers are already in extensive use for the very pre-utopian purpose of holding a precarious status quo. Even in this unglamorous role, which may just keep up for a while with population growth, the ceaseless chemical assault on the sensitive, life-bearing upper film of the earth's crust begins to exact its price in soil and water conditions; and with a rising world population, mere subsistence will push the limits of global tolerance long before utopia can show its face. I forbear to recount the cumulative penalties of other agrarian maximization strategies, such as irrigation (soil salinity), tilling of grasslands (erosion), deforestation (oxygen depletion). Sticking to Bloch's fertilizers, one reminder is crucial: that they are *forms of energy* and as such come under the double problem of the gaining and using of free energy in the closed system of the planet.

With this categorization, we also can dispose summarily of the question of the sufficiency of the mineral resources required for a permanent mass civilization of plenty. Even granted their inexhaustible crustal presence as such, their extraction from ever-deeper strata and lower concentrations for the multiplied human billions of the future *that are to live in utopian style* will require an energy expenditure such as the planet can either not deliver or not bear in its biospheric impact. Forgetting therefore about all problems of scarce or nonrenewable resources on the one hand, and of environmental pollution (agrarian, industrial, urban) on the other hand, let us concentrate on the energy issue alone.

b) The energy question   Raising the living standards worldwide to those of today's developed countries involves a dizzying multiplication of planetary energy consumption, already perilously high in its present uneven distribution; and this must be multiplied once more if the Western per capita average is to be that of the planet's multiplied billions of *tomorrow!* Here lies the crux of the whole issue: its name is energy, not matter. And the energy problem is not only that of its utilizable stock of planetary sources, the *obtaining* of free energy—what we think of first today—but the biospheric effects of its *employment,* its release in the orders of magnitude here at play. These effects are unalterably decreed by the laws of thermodynamics.

Of these effects, I mention in passing only the so-called green-house effect of continued immoderate burning of fossil fuels, perhaps prevented only by their timely exhaustion. Catastrophic climatic changes could be caused by the heating up of the atmosphere, with melting of the polar ice caps, etc. Thus, the happy-go-lucky feast of a few industrial centuries could be paid for with millennia of altered terrestrial nature—not unfairly by cosmic reckoning, since those centuries would have squandered the inheritance from millions of years of past life.

But even apart from this special effect that is bound to carbon-dioxide generation by organic fuels (thus avoided by nuclear energy), there is an ultimate thermal limit to *all* plans of technologic-utopian aggrandizement of the human condition on earth, whichever kind of energy transformation is used for its running. We presently worry about waning energy supply and chafe under its existing or impending limits. But if it were unlimited, could we use it with impunity? The answer is: no. Assuming all other forms of pollution to be circumvented—chemical or radioactive, of air, land, and sea—and assuming we have licked exhaustibility of sources by unlocking "clean" and practically unlimited nuclear-fusion energy, there would still remain *thermal pollution,* the inevitable end product of all energy transactions, and therewith the question of the admissible upper limit of heating up the closed box of our terrestrial environment. The impossibility of circumventing *this* causality by any engineering trick— to have the *use* of the energy without the thermal consequences—is ultimately the same as the impossibility of constructing a *perpetuum mobile:* the inviolable law of entropy that in every work performance energy is "lost," that all energy finally degenerates to heat, and that heat is dissipated, that is, evens out with the environment to a mean value. There is no appeal from the second law of thermodynamics. Thus, if not from energy starvation, utopia can come to grief from energy gluttony.[7]

c) *Modesty of goals versus the immodesty of utopia*   This is not to dissuade from trying to develop nuclear fusion for peaceful uses. Its eventual possession could be a timely and much-needed gift in our circumstances. But the gift must be used wisely and with moderation, with the outlook of global responsibility and not of grandiose global hope. This rules out, from the start, the utopian leisure-*cum*-plenty paradise for the 30 or 20 or even only 10 billion earth population of a not-too-distant future. Where on the ascent the upper thermal limit lies, or where the danger threshold toward it begins, remains to be calculated. A new science is needed for that, which will have to deal with an enormous complexity of interdependencies. Prior to the thermal aspect, which lies far ahead, it must include the biochemical fate of soil and water, planetary oxygen economy, and so on. So long as we have not attained certainty of prediction here, and especially in view of the likely irreversibility of some of the initiated processes beyond a still undetermined point, *caution* is the better part of bravery and surely a command of responsibility: *perhaps for ever,* namely, if such a science should transcend all real capacities regarding completeness of data and more so their joint computability. Uncertainty may be our permanent fate—*which has moral consequences.*

The call to caution, that is, to "modest" goals, dissonant as it sounds to the ear of grandiose capacity, becomes a first duty precisely *because* of this grandeur of capacity. Its threat of sudden catastrophe by atomic war (where naked fear comes to the aid of reason) is surpassed by its threat of slowly gradual catastrophe, where the always next blessings of its peaceful use drown out the voice of distant caution. Not timidity, but the imperative of responsibility issues the novel call to modesty. Utopia at any rate, insofar as harnessed to material plenty—*the* immodest goal par excellence—must be renounced; not only because, if ever attained, it could not last, but more so because already the road in that direction leads to disaster.

d) *Why after proven external impossibility the internal critique of the ideal is still necessary*   At this point one might say that a scrutiny of the *intrinsic* worth and rightness of the ideal is superfluous since its realization is anyhow excluded by external fiat. But this is not enough. Even after the demonstration of external impossibility, an internal critique is still necessary. For, first, it must not be overlooked that with *one* condition the envisaged utopian state could become realizable and physically viable: with a sufficiently low or lowered number of humans! If one does not shrink from the monstrous measures required for that, an island of the blessed may well be established for the chosen few on the corpses of countless discards. I do not impute to the apostles of utopia any such horrid intentions. On the other hand, one must not underrate the lengths to which a highest good absolutely believed in may seduce its self-appointed trustees. With the honest conviction that the present state of man is rotten and only counts as a cradle for the better that is to come, even the most extreme means might appear among the options of the faithful—the more so as the dictatorship which is postulated as a means for getting there tempts *per se* to extreme measures and surely renders them possible. In short, utopian faith, if it is more than a dream (and that is its first claim in Marxist realism), seduces to fanaticism with all its propensity to ruthlessness.

Second, the *wish,* armed with the force of doctrine, can color the estimate of facts and chances; it can interpret the always remaining uncertainty of the scientific prognostications in its favor and put its bet on the margin of ignorance they leave; it can even discredit the science itself that yields those undesired prognoses as a servant of the class enemy: correct science—ours—reads the text of reality differently. No examples needed here.

But this is not all and not even the main reason why we cannot halt our scrutiny at this point. Beyond the dangers of uncritical faith there are those of disappointed faith: the dangers of despair when merely the external unattainability and not also the intrinsic error of the ideal comes to be grasped. For the ideal, if true, condemns every other state to being unworthy of man; and it is not good to go into the soberness of renunciation with hate for that with which one has to live thereafter, with contempt for the improvements this still does admit, with disbelief in the value of that to which man's condition yet enables him within his limitations. It is also not good and right, it is ethically and contemplatively injurious, to see man as cheated of his birthright by nature that brought him forth, to see the natural order as an invidious enemy that denies him his genuine humanity. In the bitterness of this wholesale metaphysical grudge one may miss "the genuine" that is open to him. Here lies our philosophic-ethical interest in examining the merits of the ideal itself. And finally, we also owe it to the vision of great and compassionate minds who were unaware of its pitiless sides.

B. Second Step: The Dream Translated into Reality, or: Of the Desirability of Utopia

A scrutiny of the utopian ideal itself—as distinct from the possibility of its realization—must deal with these two aspects: its positive human content insofar as it is at least adumbrated in the prophecies, and its negative background, namely, the contention that past history has not yet let Man appear in his truth. The background belongs to the ideal because it prescribes to it that its human fruit is not to be conceived in the image of previous humanity but will be something qualitatively new. Though, on pain of absurdity, this cannot be taken literally, the expectation thus radically expressed colors the logic and spirit of the utopistic argument. First we shall consider the positive aspect of the ideal, its anticipated content, in terms of human life.

1. *The Realm of Freedom in Karl Marx*

We begin with a famous saying of Marx in volume III of *Das Kapital:* "The realm of freedom begins indeed only at the point where labor determined by need and external expediency ceases. It lies therefore by the very nature of the matter beyond the sphere of material production as such."[8] The cues here are "freedom" and "labor," namely so that the first freedom which the kingdom grants is that from work for external ends, these alone being the occasion for the *necessity* to work. With the liberation from this necessity, all the positive freedoms can begin to flourish, and their ultimate object, according to another saying by Marx, is "the unfolding of the wealth of human nature." This unfolding (elsewhere called "unchaining") can therefore only take place in *leisure*—not occasional leisure in the intervals of toil, but leisure as a permanent, at least prevalent, form of life. Toil reduced to a minimum through automation, manual labor largely changed to supervisory and regulatory functions— these are envisioned by Marx as liberating in the social context.[9] He does foresee, in a cryptic saying, that then "work," from a means to life, will become itself *a first need of life:* then namely when, "in a higher phase of communist society, the enslaving subjection of the individuals under the division of labor, and therewith the antithesis of physical and mental work, will have disappeared."[10] Thus a certain problem of leisure—of all that time set free for all by socialized technology—is at least hinted at by Marx, but what is mainly stressed is the blessing for "the artistic, scientific, etc., development of the individuals."[11] We are even told that the accumulated knowledge of society will then exist in the heads of those individuals![12] But let us take a look at the utopian status of work.

How does work *per se*—not its product—become "a first need of life"? Answer: by having been taken away from man by the machine, whose work secures the fulfillment of his formerly "first" needs. Thus, since man must be occupied and time must be filled, we have the paradox that the new need for work—not for its yield—becomes perhaps the hardest to satisfy of all those needs whose satisfaction the new society is mandated to assure to each of its individuals. The problem, beginning to appear already in present capitalist societies, will be all-pervasive in the utopian society that is committed, for the sake of abundance plus leisure, to extremes of productivity by means of extreme automation of production. But—taking off here from Marx—the general need to work will be differentiated, as to kind and extent, by personal dispositions. It will be the need of different abilities or inclinations to actualize themselves beyond utility, the latter being taken care of by the labor-saving, almost unstaffed robot apparatus. The need for work as such becomes a therapeutic rather than economic issue. But if, paradoxically, work itself becomes a major commodity to be distributed by society, and if the cause of general happiness requires its distribution according to individual needs thereto, then the famous Marxian formula, "From each according to his abilities, to each according to his needs," takes on the unexpected meaning: From each according to his ability to feel a need for work, to each according to the need of his ability. This would then become a principal social program, and a very extravagant one because the use value of the work and its products would not count.

Marx himself seems not to have had misgivings about the paradoxes of technologically donated or inflicted leisure. His faith in the intrinsic wealth of human nature, bound to burst forth once freedom from toil has taken the lid off it, is axiomatic. So is the vital alliance of socialism with advancing technology as the premise of its utopian outlook. No superannuation of man through loss of his work roles to the machine is feared. "Time on our hands" presents apparently no problem to the humanist in Marx. What he says about the overcoming of the enslaving division of labor and the vanishing antithesis of manual and mental work in the still-remaining human chores of socialized industry sounds rather oracular. The actual trend looks the opposite: rising rather than declining specialization seems to be the lot of all those who will still participate actively, as experts, in the running of the productive apparatus. The others—necessarily the great majority—are not so much freed as excluded from useful work. This nightmare of present-day economists and sociologists in looking at coming automation did not yet darken the horizon at Marx's time.

For more concrete thoughts on utopian leisure and its possible human content we must turn to Ernst Bloch, the grandmaster and *enfant terrible* of Marxist utopians in our century.

2. *Ernst Bloch and the Earthly Paradise of Active Leisure*

Unabashedly Bloch cites the "waking dream of a perfect life" (P.H. 1616), the "dream of the regnum humanum" (P.H. 1619), the "primal intention of a golden age" (P.H. 1621), the "absolute goal concept" (P.H. 1628) as the ultimate moving force behind the Marxist commitment. The closing sentence of his colossal work *Das Prinzip Hoffnung* calls it "something which shines into the childhood of everyone and where no one yet was: home." This is essentially more and different than justice, kindness, compassion, even love, or any other caring for the human lot down here, all of which can do their work in the world without such an expectation and never do it for its sake. The Marxist goal, according to Bloch, is nothing less than utopia, to which justice, etc., can clear the path, but in whose attained presence the old virtues, responding as they do to *negative* conditions, have little room. Now, whatever content those superlative terms—perfect life, kingdom of man, golden age, absolute goal, final home— may connote, their clearly stated premise and locus is *leisure* as a universal possession. The requisite physical premise of that again is the "reconstruction" of planetary nature, of which I have critically spoken before. Now we take a critical look at the proposition of leisure itself, imagining its physical condition supplied.

A lengthy section of Bloch's book bears the title "Leisure as an indispensable, still only half-explored goal." From it I quote: "Vanish will the differences between manual and cerebral work, between city and country, but most of all, as far as possible, those between work and leisure" (P.H. 1080). "Society, standing itself beyond labor, will for this very reason no longer have separate Sundays and holidays; *hobby will become vocation* . . . and in a happy marriage with the mind, the weekdays of this society will themselves be festive" (1071 f.).

Let us pass over the "happy marriage with the mind," which must somehow be related to the mysterious (to me unintelligible) disappearance of the difference between the work of hands and head. Let us focus on the hobby as vocation.

Unlike Marx, Bloch explicitly recognized a problem in pervasive leisure—indeed, *the* problem of utopia, once its material conditions are realized. He frankly speaks of the "utterly naked question of leisure," namely, "the at last clearly manifest problem of its ever more concrete *contents*," to which "a human answer" must be found. It will be found, he says, with the aid of the teachers who will rule and guide the classless society (instead of present-day state governments) (P.H. 1086). The "new needs of leisure itself" will, as is the way of needs, "produce a new superstructure," their own "ideology" concerning those human contents (P.H. 1083)—presumably sanctioned by the authority of the "teachers." We will be told what are worthy leisure occupations. Here we begin to feel a chill.

Anyway, the real answers to the question of leisure must wait for the perfected mankind and for its teachers to supply them. But some formal properties of its humanly worthy content can be stated in advance, and the first to Bloch is, reasonably enough, that the happiness of utopian existence *cannot be passive but must be active*—thus, just as Aristotle might have said: active leisure, not idleness. As to the manner of activity, bourgeois society already furnishes a provisional idea of it in the role of *hobbies*. These are nowadays, Bloch says, an escape from the dreariness of unchosen jobs, therefore especially widespread in America where most people do not find fulfillment in their daily occupations. But mere sideline hobbyism will cease when the hobby becomes the real vocation itself. "Until then, we can learn from the hobby how actively filled leisure is privately dreamed about: as work that feels like leisure" (P.H. 1061). So far Bloch. Let us see whether the hobby as vocation, as principal occupation, can really fill the bill, and whether in general the gourmet standpoint it represents is psychosocially viable. The question is important beyond Bloch's musings, because with or without Marxist utopia *technology* as such is heading that way with its displacement of human labor.

I can discern three deadly costs of the alleged hobby solution to the problem of leisure: loss of *spontaneity* in the hobby becoming a duty, loss of *freedom* in its necessary public supervision, loss of *reality* in its fictitious character. The last is the most serious; and the fundamental error of the whole conception, already in Marx, is *the separation of the realm of freedom from the realm of necessity,* the belief that the one begins where the other ends, that freedom lies somewhere beyond necessity instead of in the meeting with it. I very briefly touch on a few points of these critical considerations before dwelling on the last one.

a) Loss of spontaneity   One of the attractive traits of a genuine hobby is that it can be indulged on the side and half-seriously. Its choice need not, and rarely does, imply that one prefers it to one's main occupation and would have chosen it instead, if only allowed to. One look at the "hobby" entries in *Who's Who* makes that clear. A personal memory sticks with me. As a novice on these shores I sometimes met at parties a chemistry professor whose conversational obsession with his rose-growing made me observe to my wife that his chemistry and his devotion to it couldn't amount to much. Some years later he received the Nobel prize, and not for rose-growing. Admittedly, that happy coincidence of labor and love which the scholar enjoys is rare; but I doubt that even the machinist whom circumstances rather than love have gotten into the job would wish to collect butterflies all the time. When left with nothing but the hobby, the latter cannot keep its freshness and may soon take on a desperate quality.

b) Loss of freedom   More important is the *loss of freedom*. As a broad social concern, full-time hobbyism must come under public control. In the first place, the state must insist on everyone having a full-time hobby and monitor its being practiced, for idleness breeds lawlessness. Self-motivation cannot be trusted in so serious a matter. In the second place, the public purse which pays for the expenses that most hobbies involve must claim a more specific control over them. Remember that their creations add little to the main social product, this being generated by the selfsame robot technology which created the leisure vacuum in the first place. Thus the authority underwriting the supererogatory kilns, looms, lathes, etc., has to take a hand in assigning and denying hobbies according to some tolerable overall distribution. Whim goes out and certification comes in. Imagination shrinks from the specter of the bureaucracy administering such a system—with its counselors and supervisors, questionnaires, tests, psychological profiles, periodic checks, etc. In short, utopia which donates the leisure must also govern it with a paternalistic hand.

c) *Loss of reality*   But the real rub is that all this society-wide stimulation, organization, and enforcement of leisure-time activity will not help because it is mere pretending. This brings us to the cardinal point: the *loss of reality* and therewith of human *dignity*. No ideology can conceal from the actors of this scenario the dismal truth that what they are doing does not really matter, that it could as well be left undone or postponed or done sloppily, without any other damage than setting a bad example and any other penalty than a bad social mark. The ghostliness of unreality descends on the whole make-believe activity, and with it an unimaginable *taedium vitae* whose first victim must be the pleasure in even the self-chosen hobby. No serious person can be happy in a life of constant and transparent pretense. Maybe only a sensitive minority would feel it keenly, but the fictitiousness of existence must be demoralizing for all. For with realness it also takes away from man the feeling of his worth, which is what gives him dignity in his own eyes. Contentment with that condition, bolstered by a full stomach, would thus be the contentment of indignity. He who cares today for the dignity of man should fear rather than wish such contentment for future generations.

But Bloch sees matters exactly the opposite way. To him, the worries of material existence are an insult to human dignity, and the care for acquisition in their service is the shabbiest of all cares. The more society is put right economically, the more can the true cares of existence come to the fore and be attended to, those alone worthy of man (P.H. 1072, 1083). Leaving undetermined what the true cares *of* existence are, let us just stop to consider one absurd inference from the alleged shabbiness and indignity of men's struggle *for* existence itself, with its compulsion to work for a living. The Eskimo who wrests sustenance for himself and his kin from the Artic ice, as necessity bids him—does he live unworthily? And he who doesn't need to, but who can afford it and does it for sport or fun or to prove himself or to gain fame—does he do something worthier of human dignity? The hardy Phoenician sailors of old, the codfishers of Kipling's *Captains Courageous*—were they shabby in their arduous quest for gain? And the yachtsman, free from the drive of necessity, attends to the true cares of existence, one of which is what to do with one's free time? But this is absurd. And it exposes the fallacy in Marx's dictum that the realm of freedom begins where labor for external needs ends.

The truth is quite the opposite. Freedom consists and lives in pitting itself against necessity. Separated from it, freedom loses its object and becomes as empty as force without resistance. Empty freedom, like empty force, cancels itself. With the seriousness of reality, which is always necessity as well, vanishes the dignity that distinguishes man precisely in his *relation* to the real and the necessary. Play as a life's occupation, far from releasing human dignity, excludes it. Or: *there is no realm of freedom outside that of necessity!* Thus, both freedom and dignity are the losers in the hobby-busy leisure of utopia. Apart from this moral angle, it must fail even in plain practical-psychological terms: fictitious activity protects as little from anomy and despair as does chronic inactivity—and this can count almost as a consolation for the sake of man.

C. The Negative Background of the Dream, or: On the Provisional Nature of all Previous History

1. *Why after Refuting the Image of the Future, We Still Must Discuss the Image of the Past*

With this, the critique of utopia in itself could end, were it not for its second aspect, logically even its first: what we have called the "negative background" of the ideal, namely, the doctrine of the "not yet," the inchoate, inauthentic, merely provisional and even distorted nature of all humanity heretofore. It is not just for completeness' sake that we must take at least a brief look at this doctrine too. For the utopian longing as such, which is at work in and ahead of any particular goal conception, feeds upon the appraisal of what was and is. And when this indeed is found to be utterly wanting, then the search for the "totally other" true shape of things must continue, even after one particular image of it has had to give way to the force of criticism. The striving will remain utopian-revolutionary—and can do so (as the recent past has shown) even without *any* image of the hoped-for, namely, in reliance on the "force of the negative" in itself, on the power of negation to give birth to the object of eventual affirmation. It also must not be overlooked that Bloch could reply to all of our substantive criticisms that they were extrapolated from man as we know him, the "old Adam," while Marxism expects from the postrevolutionary classless society a new man, whose behavior and being is not to be inferred from the psychology of the old man. Now, one can rightly dismiss such an expectation or hope as utterly groundless and no better than the crassest faith in miracles. But this still leaves the fact that a certain view of the past has urged such a dream of the future and will continue to do so. And since we all must enter the future with the image of the past, it is indeed of significance, beyond the tenability of particular "Latter Day" conceptions, whether in that past we do or do not yet find the human essence which is to be our concern in the future as well. The ethics of responsibility itself thus requires an examination of the thesis of the "Not Yet" of all previous history.

2. *Ernst Bloch's Ontology of the "Not Yet"*

How can one arrive at the view, and what does it mean, that man as he can and "should" be has never yet been and is still to come? To take first what it does not mean: it means nothing in the way of civilizational progress, according to which indeed much (though not necessarily the "essential") is still to come, for that would merely point to a firm continuation of past development, not to a negating and innovating revolution; and it would anyway not concern the intrinsic being of individuals, the "nature" of man, but rather the external instrumentalities and collective orderings of his existence. These latter do have an influence on the "dispersion" (its width or narrowness) of higher culture within the respective societies, but our issue here is not the frequency or rarity of fulfilled lives, or the proportional size of the elites here and there in history, or the numerical distribution of outstanding individuals, but rather whether even these already represent "true man," that is, *whether he has ever yet existed at all*. And here, radical utopianism answers: no; but, as a dream: yes. In Bloch this view finds expression in a complete *ontology*, which in its philosophical scope goes far beyond man and includes the whole of being, even that of matter itself, but which we consider only with respect to man and history: the ontology of "being-not-yet."[13] Its terse formula runs: "*S* is not yet *P*" (the subject is not yet its predicate), where being-*P* is something *S* not only *can* but should attain so as to be *really S*. As long as it is not *P*, it is not yet itself (this is the "no"). The logical formula is ontologically undergirded by the concept of "latent tendency," according to which there lives in *S* an inner longing for this self-realization, for that *P:* thus a hidden teleology (this is the dream, with its "yes"). "Future" is the key word in the provisional status, in the "not yet," of all previous history.

3. *The Hidden Utopianism of Past History*

This future is indeed, according to Bloch, foreshadowed or "prefigured" in the past, and not only negatively, by its failings, but in some luminous, prophetic instances also positively. These somehow, in unconscious anticipation, express and keep alive the "dream" and are thus the only worthwhile "cultural heritage" that will even survive into the age of consummation: they embody the secret "utopianism" of all past history.

> Only the moulding dream-force for a better world or the utopian function, as a self-transcending one, is each time and always

culturally creative. This function . . . alone constitutes the sub-
strate of the cultural heritage. [P.H. 179]

To me, it is this select praise rather than the prevalent blame of past
history (the right to which cannot be denied to the honest pessimist and
moralist) that is the really objectionable aspect of the utopian "not yet."
One example from the scattered (always rather oracular) allusions to such
"prefigurations" must here suffice—one taken from the sphere of art:
"Empires come and go; a good verse remains *and tells what—lies ahead*"
(P.H. 1072, italics mine). But no! Not in the least! May I, for once, counter
this appraising tribute to great art with a personal memory of mine? When
I found myself, unexpectedly, standing before Giovanni Bellini's Madonna
triptych in the sacristy of St. Zaccaria in Venice, I was overcome by the
feeling: here had been a moment of perfection, and I am allowed to see
it. Eons had conspired toward that moment, and in eons it would not
return if left unseized: the moment when, in a fleeting "balance of colossal
forces,"[14] the All seems to pause for the length of a heartbeat to allow a
supreme reconciliation of its contradictions in a work of man. And what
this work of man holds fast is absolute *presence* in itself—no past, no
future, no promise, no succession, whether better or worse, not a prefig-
uration of anything, but rather timeless shining in itself. *That* is the "utopia" beyond every "not yet," scattered moments of eternity in the flux
of time. But they are a rare gift, and we should not forget over them the
great tormented souls, to whom we owe perhaps even more (and something other than instruction about a "not yet"): in *them*, too, there is the
ageless *present* of man. That there are yet things to come is indeed always
part of what is and each time our task, but to read it into the testimony
of the past for our benefit and edification, as if only we at last could lead
it in us beyond itself and to its destination, as if it had waited for us, nay,
had been "meant" for us in the first place—that is to rob it of its inherent
own right, and ourselves of its true gift.

4. *Our Answer to the Ontology of "Not Yet"*
Here I break off. Much more could be said about the judgment on all
of past history that it has not yet revealed what "man" truly is and can
be (a judgment in which such antipodes as Nietzsche and Marx concur).
Instead, I will merely summarize my arguments against this verdict and
its underlying ontology of the "not yet" and the "yet to come."

The basic error of the ontology of "not yet" and its eschatological hope
is repudiated by the plain truth—ground for neither jubilation nor dejec-
tion—that genuine man is always already there and was there throughout
known history: in his heights and his depths, his greatness and wretchedness, his bliss and torment, his justice and his guilt—in short, in all the
ambiguity that is inseparable from his humanity. Wishing to abolish this

constitutive ambiguity is wishing to abolish man in his unfathomable free-dom. In virtue of that freedom and the uniqueness of each of its situations, man will indeed be always new and different from all before him, but never more "genuine." Never also exempt from the intrinsic perils, moral and other, that belong to his genuineness. The really unambiguous man of utopia can only be the flattened, behaviorally conditioned homunculus of futuristic psychological engineering. This is today one of the things we have reason to *fear* of the future. *Hope* we should, quite contrary to the utopian hope, that in future, too, every contentment will breed its dis-content, every having its desire, every resting its unrest, every liberty its temptation—even every happiness its unhappiness. It is perhaps the only certainty we have about the human heart that it will not disappoint us in this expectation.

But as regards the much-needed improvement of *conditions* for much or all of mankind, it is vitally necessary to *unhook the demands of justice, charity, and reason from the bait of utopia.* For their own sakes, neither pessimistically nor optimistically, but realistically, should we heed these demands, unintoxicated by excessive expectations, thus also untempted to an excessive price, which any secular millennialism, "totalitarian" by its nature, is ready to make those pay who live in the shadow of its supposed advent.

III. From the Critique of Utopia to the Ethics of Responsibility

1. Hope, Fear and Modesty

At the end, we return to technology. In fact, we have been with it all the time. Marxist utopia, involving the fullest use of supertechnology, served as an "eschatologically" radicalized version of what the worldwide technological impetus of our civilization is moving toward anyway. Thus the critique of utopia is implicitly a critique of technology in the antici-pation of its extreme possibilities. Much of what we tried to depict as the concrete human condition in the realized dream, not counting the vetoes from the abused environment, has appeared as a cause for fear rather than hope. This, by the way, is the reason for the new literary genre of "negative utopias," which (at least since Aldous Huxley) have tended to replace the ebullient ones of former times. At any rate, the starry-eyed ethics of perfectibility has to give way to the sterner one of responsibility. The latter is not devoid of hope, but gives also fear its rightful place. Its heart is veneration for the image of man, turning into trembling concern for its vulnerability. Promethean immodesty—and utopia is *the* immodest goal par excellence—must yield to the modesty of goals that we and nature can afford. The warning lights of various limits are coming on. The time for the headlong race of progress is over, not of course for guarded progress itself. Humbled we may feel, but not humiliated. Man's mandate remains exacting enough outside of paradise. To preserve the integrity of his essence, which implies that of his natural environment; to save this trust unstunted through the perils of the times, mostly the perils of his own overmighty deeds—this is not a utopian goal, but not so very modest a task of responsibility for the future of man on earth.

2. About the Spectrum of Fears

Let us also, at the end of our long road, return once more to the "heuristics of fear" which I have been suggesting in Chapter 2. For many, the apocalyptic potential of our technology is concentrated in the atom bomb. I am sure that they do not exaggerate the peril. But it has one consolation: it lies in the realm of arbitrary choice. Certain acts of certain actors can bring about the catastrophe—but they can also remain undone. Nuclear weapons can even be abolished without this requiring all of modern existence to change. (The prospect is admittedly small.) Anyway, decisions still play a role—and in these: fear. Not that this can be trusted; but we *can,* in principle, be *lucky* because the use is not *necessary* in principle, that is, not impelled by the production of the thing as such (which rather aims at obviating the necessity of its use).

My main fear rather relates to the apocalypse threatening from the nature of the unintended dynamics of technical civilization as such, inherent in its structure, whereto it drifts willy-nilly and with exponential acceleration: the apocalypse of the "too much," with exhaustion, pollution, desolation of the planet. Here the credible extrapolations are frightening and the calculable time spans shrink at a frenzied pace. Here averting the disaster asks for a revocation of the whole life-style, even of the very principle of the advanced industrial societies, and will hurt an endless number of interests (the habit interests of all!). It thus will be much more difficult than the prevention of nuclear destruction, which after all is possible without decisive interference with the general conditions of our technological existence. Most of all, the one apocalypse is almost bound to come by the logic of present trends that positively forge ahead toward it; the other is only a terrible contingency which may or may not happen.

Therefore, with all respect for the threat of sudden destruction by the atom bomb, I put the threat of the slow incremental opposite, overpopulation and all the other "too much," in the forefront of my fears. That time bomb, whose ticking so far cannot be checked, competes in destructive power, alas, with any amount of hydrogen bombs. The apocalypse which threatens here from a total *development* (not just a single act) seems to me not smaller than the sudden one of an atomic holocaust, its consequences possibly as irreversible, and to its coming every one of us contributes by mere membership in modern society. This apocalypse waits for our grandchildren, if we are lucky enough till then to have avoided the nuclear peril.

Darkest of all is, of course, the possibility that one will lead to the other; that in the global mass misery of a failing biosphere where "to have or have not" turns into "to be or not to be" for whole populations and "everyone for himself" becomes the common battle cry, one or the other desperate side will, in the fight for dwindling resources, resort to the *ultima ratio* of atomic war—that is, will be driven to it. Probably the bomb (in a "primitive" form) will become widely enough available so that not only poor nations but even small terror groups will be able to operate with it. The apocalypse "with a whimper" can be preceded by several small "bangs," even if we will be spared the "big bang." *Both* must be prevented. Where the greater cause for pessimism is may be moot. For me personally only the overall drift with its entirely nondiabolic automatism has the traits of an almost inevitable *fatum,* which the sudden murder of mankind has not. But over that point there need be no quarrel.

3. Answering the Charge of "Anti-Technologism"

One objection has surely been waiting to be heard. Have we not over-emphasized the *threat* of technology and underplayed its *promise?* Only the voice of warning and caution has been heard, not that of an inspiring task. A heuristics of fear, so one will say, has its points, but only in counterbalance to a heuristics of hope, which has hitherto lighted mankind's path. When speaking of an ethics of technology and of imperatives in connection with it—must we not also, and perhaps in the first place, speak of an imperative on its side? In short, the objection of antitechnological bias can be raised—and then, by a slight extension, perhaps even the charge of antiscientism.

I concede the charge of one-sided emphasis, which I will defend, but deny the bias against technology as such, let alone against science. The "technological imperative" is nowhere questioned, as it is indeed unquestionable in its anthropological primacy and integral to the human condition. But it needs no advocates in the Western world of the twentieth century: intoxication has taken its place. As things are with us, the technological drive takes care of itself—no less through the pressure of its self-created necessities than through the lure of its promises, the short-term rewards of each step, and not least through its feedback-coupling with the progress of science. There are times when the drive needs moral encouragement, when hope and daring rather than fear and caution should lead. Ours is not one of them. In the headlong rush, the perils of excess become uppermost. This necessitates an ethical emphasis which, we hope, is as temporary as the condition it has to counteract. But there is also a timeless precedence of "thou shall *not*" over "thou shalt" in ethics. Warning off from evil has always, for reasons dealt with before, been

more urgent and peremptory than the positive "thou shalt" with its dis-
putable concepts of moral perfection. To keep free of guilt comes first in
moral duties, the more so when the temptations to guilt become more
powerful. Our particular version of this emphasis is in answer to the
particular, epochal, perhaps passing phase of civilization and its particular,
overpowering temptations. In thus paying tribute to the present state of
things, our one-sidedness follows the ancient ethical council of Aristotle
that, in the pursuit of virtue as the "mean" between the two extremes of
excess and deficiency, one should fight that fault more which one is more
prone to and therefore more likely to commit, and rather lean over in the
opposite direction, toward the side less favored by inclination or circumstances. This, in a time of one-sided pressures and mounting risks, is the
side of moderation and circumspection, of "beware!" and "preserve!"

Appendix to Chapter 3 (III.5)
Impotence or Power of Subjectivity

A Reappraisal of the Psychophysical Problem

Historical Prologue

I begin with an episode from the history of science. In 1845, three young, enthusiastic physiologists, pupils of the great Johannes Müller in Berlin, made a formal pact to fight the view, hitherto regnant and called "vitalism," that life involves forces other than those found in the interaction of inorganic bodies. They were Emil du Bois-Reymond, Ernst Brücke, and Hermann von Helmholtz—all to become very famous. Du Bois-Reymond and Brücke between them even pledged with a solemn oath that they would establish and compel acceptance of this truth: "No other forces than the common physical chemical ones are active within the organism."[1]

All three kept their vow throughout their lives, with spectacular scientific success, which in turn helped to make their proposition an article of faith, and "vitalism" a discredited cause, throughout the life sciences. What escaped them was the fact that by making *a* pledge, no matter which, they already contradicted or acted counter to the particular content of *this* pledge. For they did not bind themselves to what is not a matter for decision at all—namely, to let the molecules of their brains now and hereafter take their causally prescribed courses and allow them to determine their thought and speech (if they do that anyway); but they bound themselves to remain faithful in the future to a present insight, thus by implication declaring at least *their* subjectivity to be master over their conduct. In the mere fact of a vow, they credited something totally *non*physical, their relation to truth, with just that *power* over their overt behavior which the *content* of their vow on principle denied. Making a promise, with faith in the ability to keep it and the equally implied alternative of *not* keeping it, does admit into the chart of reality—if not into scientific discourse—a force "other than those found in the interaction of inorganic bodies." "Fidelity" would be such a force, as every other rule of mind over behavior, that is, body. And yet they were right to exclude the psyche from scientific discourse in their program for physiology. This is the genuine antinomy of the situation, which no one can evade.

As to the deuced psyche itself, or consciousness, or subjectivity, which has the gall to be there and is even less deniable than the being-there of bodies in space, and what to make of its copresence and interrelation with the physical basis, one of the three signers of the pact, du Bois-Reymond, coined twenty-seven years later in a famed lecture "On the Limits of the Knowledge of Nature" the much-maligned phrase that we not only do not know (*ignoramus*), but on principle never shall know (*ignorabimus*).[2] The indignation of many scientists at this "defeatism" was great, but I suspect that he was right *in the terms of natural science*. And philosophy, since Descartes, cannot boast of much better success. Still, this skeleton in the closet of modern science will insist on rattling and disturbing our sleep. Reason, as Kant perceived, cannot leave it alone—nor can it, contrary to Kant, acquiesce in a sheer antinomy. Even with the *ignorabimus* waiting implacably at the end, we might gain some ground from *complete* ignorance by at least clearing the issue of distorting and obstructing misconceptions—and in the process perhaps even achieve something better than the jarring clash of incompatibles, which after all is itself a positive assertion and as such can be in error.

I shall first try to articulate the hoary "psychophysical problem," then discuss the "epiphenomenon" formula with which it has been bent one-sidedly to the purported demands of natural science or materialism, and last I shall try my hand at a conjecture which better fits the bifurcated data of our experience, neither of which ought to be disenfranchised in deference to the other.

Statement of the Psychophysical Problem: Truth or Fraud of Consciousness

Subjectivity exists. It either is what it claims to be, or it enacts a stage play behind which another type of happening hides. In the first case, its testimony—for example, that I raise my arm because I will it—is credible at face value; in the second case, it is deceptive, a mere disguise or window dressing of neurophysiological processes, which parade in the fancy dress of will but lift the arm without will or the cooperation of will, that is, they do so irrespective of the presence of a "willing" sensation. The standpoint which grants the psyche its effectiveness stays in agreement with its self-testimony and needs no further reasons; the standpoint which denies its claims must have special reasons for doing so. These reasons, whatever their strength, cannot silence the disputed testimony itself, and thus the natural standpoint, constantly nourished by the subjective evidence, is never abolished in fact. But deprived of the privilege of naiveté once the question of credibility is raised, it must defend itself against those reasons, since at the serious suspicion of an illusion even its irresistibility ceases to count in its favor. On the other hand, the suspicion must indeed be serious. Thus one must first examine the reasons which here contest the validity of immediate evidence and put naiveté in the role of a theatergoer who takes the play on the stage for reality.

Clearly, only the strongest reasons count when it is a question of condemning all mental life to the status of illusion. I can see two such reasons and confine myself to them, ignoring all weaker ones. They are: (1) that any action of mind on matter is *incompatible* with the immanent completeness of physical determination, that is, that the latter does not tolerate such an interference from outside: this I call the "incompatibility argument"; and (2) that the mental as such is also *incapable* of intervention, being nothing but a unilaterally dependent concomitant of physical events and lacking any force of its own: this I call the "epiphenomenon argument." The first argues from the nature of the physical, the second from the nature of the psychical. How strong is either?

I. The Incompatibility Argument

1. *The Argument*
The proposition that the context of physical determination is "closed" and does not tolerate the intrusion of nonphysical causes follows from the rule of the laws of nature, especially the constancy laws, which would be violated each time when a causal quantity with no physical predecessor were added to the given sum or subtracted from it with no physical successor. One or the other would happen whenever in the acting of living subjects the further course of events differs from what it would be without the intervention of the psychical factor, that is, by the corporeal mechanics alone. In the physical reckoning, the intervention would amount to something emerging from nothing or vanishing into nothing, and this is excluded by the constancy rule. Ergo: there cannot be an influence of the mental (nonphysical) upon the physical; ergo: things proceed exclusively according to the physical *concatenatio causarum;* ergo: the addition of the psychical (subjective) dimension in living beings is gratuitous and redundant for the course of events; ergo: the consciousness of aims, etc. (feelings of willing and acting) is but a deceptive imagery for the causal working of the bodily mechanism— a deception not even excused by a purpose, since *ex hypothesi* the self-sufficiency of the masquerading facts needs no such help: a purposeless deceit of purpose.

2. *Critique*

a) Absolute determinism in physics—an idealization   How sound is the argument? We observe that it invokes not simply the validity of the constancy laws, which is to be taken as inductively proven, but their *unconditional* validity, that is, one impervious to exceptions, and this, of course, lies beyond inductive proof. Inviolability on principle pertains to the logical nature of mathematical, not of factual, rules. For the latter, it is merely postulated by us for the sake of the idea of lawfulness. The postulate originates in an idealization and expresses an ideal. The incompatibility, therefore, which the argument states, is of the type that "what must not be, cannot be." The force of the "must not" is proportional to the theoretical dignity of the ideal from which it issues. But since we, and not logical necessity, have invested it with that dignity, we can also reconsider and, if need be, modify it, provided we stay in agreement with observable facts.

b) The logic of the incompatibility problem   Moreover, an incompatibility of the nonlogical sort we are faced with designates as such no more than a difficulty of thought and leaves open *what* must be revised for its resolution: the concept of that which is to conform to a norm, or the norm to which it is to conform, or both? To decide this question, we must compare the strength of evidence which both have on their side, but also the *consequences* from any one side yielding to the other: in our case, by asking what becomes of "nature" if her causal purity, or her "exactness," is adulterated, and what of "mind" if its effective power is denied. The scales are tipped by which sacrifice is theoretically more insufferable, that is, more devastating for the side that is to make it. It is a question of the relative price of compatibility, if a price is to be paid. The contention I am going to argue is that on the side of "nature" (where the all-or-nothing logic is inappropriate), the required concession in deterministic rigor would not be devastating for its scientific concept; whereas the concession asked for compatibility's sake from the mental side, namely, the forfeiture of causal force, destroys its concept completely and even drags the favored physical side down into its own ruin, leaving a caricature of "nature" as such.

c) Historical overstatement of the determinist case   The seriousness of the problem lies in the challenge of materialist science to inner experience, which has on its side immediate self-certainty but no systematic predictive science; whereas natural science, with no immediate evidence for its generic ideal of objects, can produce constant heuristic confirmation for it in the systematization of phenomena. Because of this tested verification, a conflict with the ideal is a serious problem. But in the psychophysical incompatibility verdict the ideal is granted more than it is entitled to by its heuristic yield (its only evidence) and more than is necessary to preserve its rightful epistemological position. Unconditional determinism has always pretended a greater knowledge of nature than we possess and ever can possess. The holding of the constancy laws admits of degrees of rigor in detail. There is no *a priori* certainty that what holds for the whole also holds for all its parts down to the smallest; and what holds for the end result holds also for all intermediate links; and what for measurable time intervals holds also for each instant. The identical validity for every part and every instant presupposes an exactitude of nature which precludes any "more or less" and "approximately" and makes nature as pure as mathematics is. It was from mathematics that the idea of such an absolute exactness was borrowed, together with the homogeneity of whole and parts required for its application. *Vis-à-vis* nature it cannot be more than a postulate with which the order-loving human mind perhaps hits the truth of nature, perhaps overtaxes it.

II. The Epiphenomenalist Argument

Leaving the case of "nature" and its alleged causal closure in abeyance, we pass from the argument that the physical does not tolerate psychical interference to the converse one that it need not fear it since the psychical (the "subjective") is devoid of all causal force. The most complete expression of this position is the "epiphenomenon" thesis, whose reasoning runs somewhat like this:

1. *The Argument*

a) The primacy of matter There is matter without mind but not mind without matter. The first is demonstrated by all lifeless nature plus a good part of living nature; the second by the fact that all "mind" (here the blanket title for subjectivity of every kind and degree) appears only in conjunction with certain organizations of matter—organisms, nerves, brains—and that no example of bodyless mind is known. This observation suggests that matter has independent and primary being, mind only secondary being derivative from it.

Experience further teaches that matter in these forms of organization not only provides the necessary, enabling basis or precondition for mind's *existence,* and not only is the originating cause of that existence, but is also the continually determining cause for its *working* and all its changing contents—thus its necessary and sufficient cause in every respect. This is demonstrable for sense perception and certain feelings and emotions, and hence can be extended to thought processes for which it is not yet demonstrated: all these inner, subjective phenomena are wholly the effect of physical causes. But note that from this causation they do not gain a causality of their own in continuation of the former (as do effects in general), since they are nothing but an expression of what happens in the physical substratum. Mere expression cannot influence what it expresses, nor even itself, since then it would cease to be mere expression. Thus the "induced" subjective appearance cannot emancipate itself in either sense: it can as little play within itself as it can react on the "inducing" substratum. Its pure "as if"–pretense of both, which it displays to itself, has only the entertainment value of an illusion.

b) *The completeness of physical determination*   Joined to the impermeable causal completeness of the physical substratum previously argued, the epiphenomenon status of mind then compels the conclusion that in the case, for example, of raising my arm, *only* the objective, neuromuscular explanation (or description) is correct, while the subjective one—in terms of will and intention—is a nonauthentic symbolic transcription for it. Neurology, communication theory, and cybernetics are at work to implement this postulate—still largely empty with regard to the higher functions of consciousness—by concrete mechanistic explanations or model constructions for increasingly complex cerebral processes. The initial successes justify the expectation, so the protagonists say, that no barrier of principle will stop progress on this road. For the rest, they challenge the "spiritualists" to show how to represent theoretically a moving of physical entities by nonphysical movers and not play havoc in the attempt with the laws of physics.

c) *The redundancy of subjective purpose*   Among the subjective phenomena there are the so-called ends or purposes. For them, too, it must hold that their subjective presence has no possible influence on the course of things. Rather, as already their presence only mirrors an objective state of the substratum, so also the putative acting "*because* of them" ensues in truth *from* the same material conditions of which this appearance itself had been a symbol. The counterwitness of subjectivity on this point is not to be accepted. Socrates, therefore, was not right when, in the famous discourse of his last night, he rejected the corporeal explanation of his sitting there and awaiting the hemlock and proclaimed the explanation in terms of mind—his ideas of right and duty—as the only true one. Neither are those right who view the two alternatives as complementary, as equivalent and interchangeable aspects of the same reality. Only the Ionian naturalists had the right view, for only the physical description explains what physically happens.

d) *Mechanical simulation of behavior*   The test is here, as already Descartes had laid down as a rule, the possibility to *simulate* behavior by mechanical devices (automata), and this possibility has since been extended right into the mental realm, which Descartes still deemed exempt from such simulation. But then there also holds the further Cartesian principle that what is perfectly imitated is thereby no longer simulated but duplicated, and that then the duplicate discloses the *nature* of the original, to which no other principles of operation must be attributed than those needed for its duplication. Even imperfect imitation, if it is merely a matter of degree, establishes the perfect imitation as theoretically possible. Thus, if intelligent purposive behavior *can* be simulated in some simple forms, the more complex ones are *in principle* covered by the feat. Then the *ideas* of purpose, and other subjective data, which in the original concur with the behavior, are redundant for its actual performance and thus have no role in it. Thus the newly discovered possibility of imitating mind by purely corporeal means strengthens the speculative hypothesis of epiphenomenalism about the impotence of mind in general and the functional otiosity of psychological purpose in particular. Like all consciousness which attributes authorship to itself, teleological belief has the character of a purely putative, operatively redundant and inexplicable "as if."

2. Internal Critique of the Concept of "Epiphenomenon"

So far the materialist-epiphenomenalist argument. My discussion will focus on the *causal* aspect, which is the hub of the argument, and in its spirit I open the debate with an incompatibility argument of my own. Epiphenomenalism makes matter the cause of mind and mind the cause of nothing. But causal zero-value is compatible with nothing adhering to matter; and in particular it runs plainly counter to the idea of causal dependency itself that something dependent should be an end only (effect only) and not also in its turn a beginning (a cause) in the chain of determination.

a) The first ontological riddle: creation of soul from nothing   Before pursuing this line of thought, let us take a closer look at the concept of "epiphenomenon" as such. It says in general that the subjective or the psychical or the mental is the concomitant of certain physical occurrences in brains. The "concomitance" is one-sided, not reciprocal: the physical processes, as the primary reality, are autonomous; their secondary psychical expression is totally heteronomous or a mere product of an other. The presence of the product makes no difference to the history of the producing "other"; neither does the feat of production itself, which as it were happens "behind its back," on the basis of it but without a contribution from it. The product is a *by*-product of the intraphysical producing, with no expenditure deflected to its production, which thus is not a transitive act of the physical base but merely the "appearance" of its immanent functioning. This functioning goes on as it would anyway, whether or not it had occasioned such an accompaniment. Thus the occasioning of it is *causaliter* a "creation from nothing," since nothing was causally spent on it. Otherwise, the epiphenomenon hypothesis were useless, since the quantity expended would physically have disappeared and its converted psychical succession were no longer quantifiable—precisely what was to be avoided for the sake of the constancy principle. *The soul's causation from nothing is the first ontological riddle which the epiphenomenon theorem braves in deference to physics, in which otherwise never a thing is supposed to arise from nothing.*

b) *The second ontological riddle: a noneffective physical effect*   What thus has been produced from nothing must also remain a nothing, causally speaking. Just as the occasioning of the "accompaniment" must cost the occasion nothing, so also its being there must not change the happening which it accompanies. This indeed is the first concern of the epiphenomenon thesis: the monocausality of matter, to be safe from interference, demands the *impotence* of mind in the first place, then its cost-free generation in the second, if it is also to come *from* matter. Admittedly, the construction becomes at least logically consistent thereby: only that which is made from a causal null can also remain a causal null, having inherited no force from its cause. And yet it *is* something itself and not nothing, its being there clearly different from its not being there. The appearing of consciousness adds something to the composition of reality by which it becomes different—descriptively, but not dynamically: it "is," but nothing follows from it for the rest of things, nothing propagates from it into their further course—the rest of reality remains what it would also be without this mirage in its midst. That which is endowed with consciousness does not behave as it does because it is conscious but is conscious because it behaves as it does, that is, as its physical makeup makes it behave. Accordingly, the concept of consciousness is negatively determined thus: a something for itself and yet a nothing for other things, a being without consequence, a noneffective fact. *That something, itself a consequence (an effect), be barren of consequences, is the second ontological riddle which the epiphenomenon theorem braves in deference to physics, in which otherwise nothing is supposed to remain without consequences.*

c) *The metaphysical riddle: existence of a "delusion in itself"*   The no-consequence rule applies to the mental in two directions: outward toward the physical sphere, as just explained, but also inwardly toward its own continuation. That is, the impotence must, besides that of determining the body in action, also include that of self-determination of thinking in thought. Otherwise, mind would cease to be epiphenomenon and could go its own ways, on which the concordance with the body events might be lost. In that case, the illusion would explode when the mental sequence reaches the point of action, that is, of intentional determination of the body, and the impotence would be revealed: I will one thing and my arm does the other. But it is precisely the appearance of *power* which the impotence thesis is designed to "save." True appearance of the impotence would falsify the theory of impotence, which is just a theory of the deceit and not of the truth of consciousness. It is essential for the impotence thesis that the impotence remain hidden behind the appearance of power. To ensure this, the impotence must be indivisible: only with equal inward impotence can the outward impotence remain hidden. Internal power with external impotence would result in a consciousness at odds with the world. But the theory was devised to show a consciousness united with the world—first of all with the body. Since only the world, namely, matter, has genuine reality, the unity is nothing but the negation of *any* self-reality of the other side, therefore of internal as well as external causality on its part.

This is just what the concept of epiphenomenon provides for, by denoting something which each moment originates anew from the basis, and whose continuation, therefore, is not its own but that of the basis. As, in a motion picture, the next phase of a movement seen on the screen does not originate from the preceding phase, as it appears to do, but independent of it from the projector that emits both, and thus on the screen, contrary to appearance, in truth nothing moves—so also the temporal successor of a "now" of consciousness cannot come from this but like itself must come from the physical substratum of which each state of consciousness is, by definition, the epiphenomenon. Thus it "mirrors" the progress of the substratum, while appearing as progress of itself. As in the movie, this appearance is mere illusion. The hammer shown to crash down on the anvil is an image sequence, not a dynamic action; the deliberation terminating in a resolution is a sign sequence, not a real bringing-about: its dynamics too is illusory (the internal one, even before its external sequel when, e.g., as a seeming result of the inner sequence, a real hammer is swung).

The real dynamics which the sign sequence represents is the cerebral one, which expresses (and at the same time conceals) itself in the "text" written by it on the page of consciousness; but the continuity of the text is nothing but the continuity of the writing of it, which from below ever anew generates each sign of which it is composed and by no means lets the text write itself. The way goes not from sign to sign but from brain state to brain state and only hence each time to its equivalent in the sign script. To stay with the movie comparison: it is the continued projection that generates the continuation of the projected image; this itself is powerless. So also the "sign" on the "screen" of subjectivity in relation to its successor. No thought engenders further thought, no state of mind is pregnant with the next. But since precisely this is intrinsically claimed by them, it follows that all thinking is deception already in itself, and then once more when it believes to pass outside itself into bodily action. The delusion of self-determination overarches the delusion of body-determination. Both delusions are essential to the psyche. "Soul" *is* that very delusion or make-believe with which reality constantly deceives itself. Itself? But no, brains are not deceived. The subject? But this already is a deception as such. And why? Again no answer, not even that of *l'art pour l'art*, which fits Descartes' evil spirit, but not unintentionally fraudulent matter. The deceiver, whoever be the deceived, earns nothing from his deceit. The game is played to no player's benefit and no victim's harm. It cannot serve a purpose as such. "Purpose" is a cipher of something that is by nature purposeless. The script in which the cipher occurs— mentality—has itself no purpose, let alone a function. Its pageantry as a whole is a mirage. *The existence of such a "delusion in itself" is the absolute metaphysical riddle which the epiphenomenon theorem braves in deference to physics.*

There are other riddles, of a more logical kind, to be exposed in the concept of epiphenomenon, but those shown here may suffice. Riddles by themselves do not dispose of a theory, but they surely make it suspect. We now advance some arguments against it: (1) its internal inconsistency in that it violates the very concept of nature to which it pays homage, (2) its absurdity as a theory-destroying theory.

d) The causal inconsistency of the concept   The inconsistency was mentioned before: "soul" has to be ineffectual so that causal law be inviolate. But from this same law the "epiphenomenon" itself would be a singular and inexplicable exception—in coming to be at no causal cost and in existing with no causal effect. Changing roles, *we* must now defend the principle against its defenders and insist that *nothing* in the world is "for free" and nothing, once there, ends with itself, that is, leaves no issue in the world. Everything caused must itself become a cause. But the "soul," so materialism holds, belongs to the world as a result of matter. Then (*a*) matter must spend something on it, and the balance sheet of a material process must read differently, depending on whether or not it has engendered an effect in the form of consciousness. *Something* of it *must* have passed over into the effect, even if in this case we do not know the equivalences for an accounting. Something was exchanged for something. Conversely, (*b*) the existence of this new datum (the resultant copresence of mind), surely different from its nonexistence, must have a share in the progress of events, since no difference of existence is dynamically neutral—even if here again we are ignorant of the manner of transfer. Only on this condition has the expenditure for its becoming not simply vanished from the world, and the balance of the whole is preserved.

The principle is simply that as little as something can arise from nothing it can vanish into nothing. Nothing is pure beginning and nothing pure ending. Only the indeterminist can admit exceptions from the rule; but the materialist argues the cause of determinism and is bound to it. Thus his "solution" contains a real self-contradiction, namely, saving the inviolability of a principle by its violation.

3. Reductio ad Absurdum from the Consequences

So much for the internal critique of the *concept* of epiphenomenon. More devastating still are the *consequences* which flow from it for everything else: for the concept of a reality that indulges in this kind of thing, for a thinking that explains itself by it, and for itself as a thought of that thinking. Here the charge is not inconsistency but absurdity.

a) An absurd nature   First, what sort of being would that be which brings forth, as its most elaborate performance, this vain mirage? We answer: not a merely indifferent, but a positively absurd or perverse being, and therefore unbelievable. If living behavior were nothing but a deaf-mute pantomime, performed by supremely sophisticated physical systems without enjoyment of subjectivity, it could well be termed pointless but not strictly absurd. The show becomes absurd when it accompanies itself with a music *as if* its predecided paces were set by it. A lie can have a function, but not here: the mechanical needs no bribe. And yet it should sound— in will, pleasure, and pain—a siren song with no one there to seduce? A song that only sings its error to itself, including the error of being the singer? Something devoid of interest in the first place, and with no room for its intercession in the second, should stage the grandiose comedy of interest, shamming a task that is not there and a power it does not have? The sheer, senseless futility of such an elaborate hoax is enough to disqualify it as a caricature of nature. He who makes nature absurd in order to circumvent one of her riddles has passed sentence on himself and not on her and has forfeited the right to speak anymore of laws of nature.

b) A theory-destroying theory   Even more directly than via the slander of nature has he passed judgment on himself by what his thesis says about the possible validity of any thesis whatsoever and, therefore, about the validity claim of his own. Every theory, be it the most mistaken, is a tribute to the power of thought, to which in the very meaning of the theorizing act it is allowed that it can rise above the power of extramental determinations, that it can judge freely on what is given in the field of representations, that it is, first of all, capable of the *resolve* for truth, that is, the resolve to follow the guidance of insight and not the drift of fancies. But epiphenomenalism contends the impotence of thinking and therewith its *own* inability to be independent theory. Indeed, even the extreme materialist must exempt himself *qua* thinker, so that extreme materialism as a doctrine can be possible. But while even the Cretan who declares all Cretans to be liars can add, "except myself at this moment," the epiphenomenalist who has defined the *nature* of thought cannot make this addition, because he too is swallowed up in the abyss of his universal verdict.

Thus we have a twofold *reductio ad absurdum,* according to the twofold question of what to think of a reality that brings forth this futile image, and what of the attempt of this self-confessed mirage to establish a truth about that reality. Nature as an impostor on the one hand, a theory destroying itself on the other, was the outcome of the scrutiny.

III. "Epiphenomenalism" Voided by the Voiding of "Incompatibility"

But, so we ask at last, was this suicidal *tour de force* really necessary? Are the reasons prompting it indeed so compelling that they only leave this counsel of despair—or that of conceding defeat before an insoluble riddle? The latter, of course, would be the proper option if this were the choice, and there would be nothing shameful about its modesty and honesty. But I think one can do better. I think one can show to the upholders of physical causality that their cause admits a resolution of the dilemma that does not injure it on their own terms. To this end I take now the liberty to engage in a thought experiment which is not meant to produce a theory but merely to illustrate possible compatibility by a freely constructed logical model.

Tentative Solution of the Psychophysical Problem

1. *A Thought Experiment*

Let us assume that a geometrically perfect cone stands with its apex on a surface over a center of gravity which lies exactly in the prolongation of its axis, that is, this rises exactly "vertical" from its support. With perfect symmetry and complete absence of other forces or force differentials the cone would stand in absolute, but absolutely unstable, equilibrium. "Absolutely unstable" means that an infinitesimal influence would suffice to tip it over and make it fall to one particular side. Let us think it to be a giant cone: the smallest disturbance of symmetry from without or within would trigger giant consequences. These consequences in their total course, in acceleration, force of impact, mechanical and thermal effects, equalization of the initial energy gradient, would each and all be governed univocally by the laws of nature and be calculable according to them—except the accident of the direction of the whole, for which all other directions, that is, an infinity of such, were equal candidates. As to the trigger impulse that did decide the direction, its minimal value does not enter into the measurement of the consequences with their incomparably higher order of magnitude, and for purposes of *their* causal description the unknown starter x equals zero. To be sure, the universal physical hypothesis would postulate for this x again, infinitesimal as it be, that it was determined in accordance with the constancy laws; or generally that there was a "sufficient reason" in the antecedents for this one direction of tilting being selected over all the equally possible ones. But operationally, the postulate has to remain empty for lack of verifiability, and it is obvious that here, at the critical balance point, there is room for sheer randomness or indeterminacy without detriment to the strict determinacy of the movement once underway. Indeed, computationally it would make no difference if for the initial x we posited a psychic, nonphysical origin. In the present case there is no reason whatever for doing so, but this may change as we come to other examples. So far the thought experiment shows that neutral interstices in the matrix of physical determination, properly located, would not interfere with the validity of physical law and the intactness of causal "bookkeeping."

2. *The Trigger Principle in Efferent Nerve Paths*

Our first example was nothing but the simplest, crudest, and thus utterly unrealistic illustration of the *trigger principle*, which plays such a crucial role in higher organizations of matter, that is, in the kingdom of life. Let us go straight to the summit of the pyramid where we encounter the subtlest form of the "inverted cone." Assume we have the trigger points A, B, C . . . , at the primary control centers of efferent nerve paths in the brain, corresponding to the respective possible motor commands a, b, c . . . , thus representing the "yes or no" for the actions α, β, γ . . . ; and further assume a physical state in which the chances of activation are equal but alternative for all (any one, but one only, being eligible), that is, that the decision on *which* of them will be "fired" is completely in the balance; and finally assume that for the activation, that is, for occasioning the transition from potentiality to actuality ("triggering"), an influence of the smallest order is required. What then would be the situation physically if the "choice" has actually fallen on A? Well, the ensuing transaction a, that is, the neural tranmission of the "command," and then the transaction α, that is, its motor execution (and thereafter everything which follows from it in the external world), can in unbroken sequence of determination be described, and thus explained, according to the laws of nature (thus ideally also be predicted)—without there having to be an answer to the question of why A rather than B or C had been activated. Since the magnitude responsible for that has zero value for the account of what is observable thereupon, it makes no difference for the latter's consonance with the laws of nature whether A or B or C has been activated. The alternatives are equally orthodox in their physical behavior, equally possible "before" and equally deterministic "after," and only the decision on which of them is allowed to become operative is indeterministic *on this plane* of calculability.

Nevertheless, for this x too (i.e., for the initial resolution of indifference) one might in theory at least insist on a physical accounting, considering that the relative "zero" of the triggering factor is, of course, not a real zero but *some* quantity (even, according to quantum theory, of an indivisible minimum value), and one has a right to ask whence this came and what *its* prior determination was. Two answers are possible on physical terms: either its incidence on A was a purely random event in the given quantum field, or it followed deterministically from the antecedent distribution therein. Both answers are physically acceptable on principle, but both here would be in contradiction with the facts: the randomness with both the physical and mental facts, the mechanical determinism with the mental alone (as shown before), in that it would allow them no power at all.

3. *Possibility for the "Triggering" to Originate in Mind*

There is a third alternative, which admittedly would no longer be a physical explanation: that the physical quantity required for the selection of the neuron (a sort of "Maxwell's demon") is *generated* on the part of "subjectivity" or "psyche," that is, from beyond matter. Let us not be too frightened by the anathema of this idea to physical orthodoxy to examine its implications. It speaks of a *de novo* increment to the antecedent physical sum, the insertion of a value not accounted for from within the physical system, and in that respect a *creatio ex nihilo*. Its smallness would make it unspottable in any computing of physical factors for the macrosequence of events, but the effects of its trigger function in the hypersensitive balance of the threshold condition can be immense: war and peace, rise and fall of empires, building of cathedrals and atom bombs, greening or wasting of the world may be "caused" by its infinitesimal contribution. All these chains of action in their macrocourse would offer to causal analysis a deterministic picture in complete accord with the constancy laws, as *also would innumerable others:* only the prebeginning of each, which selected it from among the alternatives, would be indeterministic, that is, not physically *but mentally* determined—just *as direct experience tells us.* From the point of view of nature, the one "direction" actually taken would be (as with the cone) an absolute accident, which as such is beyond verification.

So far, so good. But an objection is obvious: many vanishing magnitudes can add up to noticeable ones; and what we have been speaking about was not a rare event, but something occurring constantly and in countless cases, namely, with every outward action of every "subjectivity." Thus, even if willing, uneasily enough, to grant the occasional happening of a sufficiently small "creation from nothing" in the context of physics, we may well balk at what our suggestion seems to amount to: that the world of life, across the whole breadth of subjectivity diffused through it, incessantly donates antientropic energy to nature, unpaid for by increased entropy elsewhere. This, apart from its repugnance to general theory, robs the hypothesis of its saving grace of being innocuous to the rule of the constancy laws, since the accumulation of the singly nonmeasurable must eventually grow into the dimension of the measurable and, hence, into visible conflict with the requirements of those laws.

4. The Dual, Passive-Active Nature of "Subjectivity"

So it would indeed, if the case as presented up to now were complete. But its reverse has yet to be added, and when complemented by it, the hypothesis by no means stipulates a one-way flux of becoming. First we simply recall that matter was supposed to be the prior donor in supporting subjectivity as such, and so it may just receive back its original outlay in the "influx" we discussed. More especially we must remember that—as afferent nerves correspond to efferent ones—consciousness in its overall world-relation is essentially a two-way street and not a one-way street. Action *into* the world, thus far considered alone, is based on information, an input *from* the world, that is, ultimately on sensibility. But in every act of sensuous affection, the physical chain terminates in a mental representation, namely, the percept in question, and this too cannot be free of causal cost: some value must have vanished from the physical ("objective") side to reappear on the mental side in the radically different form of subjectivity. If we were nothing but contemplative beings who only perceive the world, there would be constant transfer and drain from the physical order, thus the same embarrassment as before with the opposite sign; if only active beings (e.g., with an *a priori* intellectual knowledge of all objects of action), then there would be constant reverse transfer and inflow with *its* embarrassment. Since we are inseparably both, and this in essential complementarity, it is not unreasonable to assume that, in the physical average, outflow and inflow balance each other with reference to the whole phenomenon of subjectivity, and the two opposite embarrassments cancel out.

The key to a solution of the psychophysical problem, that peculiar impasse which nothing but a philosophizing physics has created for theory, lies—so we suggest—in an age-old insight which has never been utilized in this connection: that our being *qua* subjects has this double aspect and consists of receptivity and spontaneity, sensibility and understanding, feeling and willing, suffering and acting—in brief: that it is passive and active in one. To what model construct, then, does our thought experiment lead?

5. *The Speculative Model*

Using a metaphor, we say that the net of causality is widemeshed enough to let certain fish slip in and out. Or with a change of metaphor, at the "edge" of the physical dimension, marked by such peaks of organization as brains, there is a porous wall, beyond which lies another dimension and through which an osmosis takes place in both directions, with a priority of that from the physical side. What thus physically seeps out and in is of too small a magnitude to show up quantifiably in the single case and mutually so balancing in the total as not to affect the verifiable overall working of the constancy laws. In virtue of the trigger principle, the smallness of the single input or output does not preclude great physical effects. Passage through the "wall" means each time a radical transformation in kind, such that any relation of equivalency, even the very meaning of quantitative correspondence, ceases to apply. The greatest thoughts with the mightiest consequences can arise from the tiniest physical input, and the tritest just as well. What matters is that between input and output there is interposed a process of an entirely different order from the physical one. Short or long as may be the loop of the circle that passes through the mental field on the other side of the wall, it does not move by the rules of quantitative causality but by those of mental significance. "Determined" it is too, of course, but by meaning, understanding, interest, and value—in brief, according to laws of "intentionality," and this is what we mean by freedom. Its yield is eventually fed back into the physical sphere, where everybody can recognize it (for everybody knows that unthinking nature builds no cities), without any single physical nexus confessing to its share. With the transfer forth and back, where egress and ingress go on continually, the total balance for the physical side remains even (nothing analogous applies to the mental side), and it is on the plane of that balance that natural science does its explaining. The *understanding* of the same event is done from the plane of that which for the moment stands outside the balance and is "transcendent" to it in this sense. In that understanding, the extraphysical interlude is recognized as the true origin of the physical action, though only infinitesimally its "cause."

For mentality, thus, the brief formula holds: generated by minima of energy, it also can regenerate minima of energy. In between, these minima are gone from the physical surface, yet have not vanished into real nothing anymore than has a subterranean river; thus, they also do not emerge from a real nothing when consciousness acts back into the world. The "in-between" itself is the realm of subjectivity and (relative) freedom.

Accordingly, the brain is an organ of freedom, but precisely on condition that it is an organ of subjectivity. To put it the other way around: supposing a brain of the same physical constitution as the human, but without concomitant subjectivity—we contend that it would not produce the same effects in the visible world (though perhaps quite respectable ones in body control) which we know the human brain to do. That is to say, subjectivity is not a causal superfluity. No more need to be said here about the difference in the "intelligence" of machines and of the human mind. Actually it is my view that the hypothesis of a "merely physical brain" (except in a corpse) is inadmissible. Such a physical organization *eo ipso* means in its functioning the opening up and sustaining of a psychic dimension, which then participates in the overall causality of the system with the leverage of its key position. Obviously the freedom thus established is not absolute but confined to the latitude which physical necessity itself allows it. It does allow, as we have illustrated, that the smallest force can wield the greatest power when in the given "critical" configuration it suffices to "tip the cone."[3]

That there are, in recurrent readiness, situational sets of such poised "cones" and thereby sets of options among physically equivalent possibilities is the functional meaning of such physical organizations as brains in control of organisms. Riding on the crest of this physical organization— one of whose roles is that of amplifier—"mind" with its immeasurably small physical input can be the initiator and determinant of physical effects in that order of magnitude in which visible behavior takes place.

Ontologically it should be noted that, according to this model, the "beyond" of the dividing wall or membrane is not a no-man's-land which keeps its inhabitants for itself and in which they can lose themselves as in a spirit world. Just as it only lives on the continuous input from the physical side, so it feeds back into it what has gone through the transformation of subjectivity. Mind thus belongs to the one and same ontic reality as matter, only with a thoroughly different nexus of ontologically different elements within its own dimension. In other words—as the voice of the "self" has always been telling us—the one, coherent, convertible, and intercommunicating Being is not exhausted by its massively prevalent physical aspect.

6. *Evaluation of the Model*

The model we have constructed is admittedly crude. We need not bother with whatever refinements it is susceptible of, since it does not even claim to be "true," that is, to portray what is actually the case. It is a mere play of thought, meant to illustrate that *on the "physicalist" premises themselves* the psychophysical impasse of their making (and their alone) logically admits of better solutions than the wholly unacceptable one of epiphenomenalism. As a point of strategy, not of conviction, we have made maximum concession to the materialist case and its conception of the nature of matter. The truth, I suspect, would look vastly different—not only more subtle but also framed in ontological terms which would alter our very speech of "matter" and "mind."[4] For our purpose it was enough to come up with a hypothetical fiction in the conventional terms that satisfies these three requirements: to be self-consistent, to be consistent with observable facts, and to spare nature and ourselves the scandals which materialist epiphenomenalism has been shown to inflict on both—on nature the scandal of an alleged procreation of mind without causal cost and consequence, on mind the scandal of utter futility in both action and thought. We have at least recovered, even for the captives of physicalist creed, a good intellectual conscience for believing in the immediate evidence of our being.

IV. Quantum-Mechanical Review of the Proposed Solution

The first publication of the "tentative solution" to the psychophysical problem, in 1976, had one highly welcome result in that it enticed Professor Kurt Friedrichs of the Courant Institute of Mathematical Sciences of New York University to volunteer his critique and constructive comment. In many hours of discussion over repeated sessions, he reviewed the issue and explained that a "model" answering it—if any can be found—must be framed in terms of quantum theory and not of classical physics. I shall try to communicate the instruction thus received as it bears on the thought experiment I had initially proposed. The shared premise was that "epiphenomenalism" is indeed untenable, and a better alternative to it must be found. Professor Friedrichs's observations were meant to help the search for it. I gratefully acknowledge his much-needed and so generously given aid. Needless to say, the responsibility for what follows is still mine—a layman in these matters.

1. The Incompatibility Argument: Valid on the Plane of Classical Physics

As to the present inadequacy of the model, the objection was not so much to the inelegant idea of the mind swallowing and regurgitating physical magnitudes that have meanwhile vanished from the physical scene, with the (ever so slightly and momentary) blurring of the laws of classical physics this entails, as to the attempt in general to wrest any latitude for an indeterminacy of any kind from classical mechanics, to which my model remained conceptually wedded. Prequantum physics, it was stressed, cannot deal with the problem of mind except in a completely determinist manner and is thus truly not compatible with an interaction between mind and physical processes, if mind is something other than physical process. Of particular relevance to our problem is the principle of action and counteraction (force-counterforce), without which no causal transaction is conceivable. But the subjective sense datum, for example (as distinct from the objective physiology of sensing), granted its being in some sense an "effect" of a physical cause, does not in turn *react* on it, or together with it constitute a case of "counteraction," and in the incommensurability of the mental and physical as such, a dynamic correspondence of that kind (whichever way the traffic goes) has simply no place. There is no real interaction. In sum: prior to the advent of quantum physics, the incompatibility argument was right.

2. *Indivisibility of Inner and Outer Power of Subjectivity*

With regard to the *selection* of alternative physical possibilities by mental choice, note should be taken of two different meanings thereof: (1) purely mental: sitting in my armchair, I select Carter as my candidate, that is, I "make up my mind" without at this time performing an "action"; (2) I elect Carter by pulling a lever in the voting booth, that is, out of my mental determination I perform a *physical action*. The manifest locus for the psychophysical problem is (2), and for the physicist the only one, though it was shown in the original argument (II.2.c) in this Appendix, that already (1), that is, the self-determination of mind in thought, does raise the problem with its claim to autonomy *vis-à-vis* the physical realm, and that ultimately "power" or "impotence" of mind holds either in both respects or in neither. Considering that thought itself has, or depends on, a physical organ, the brain, one can even argue that the two cases differ only as intracerebral activity and brain-transcending motor activity, both being equally physical processes, one on the microscale, the other on a macroscale. But we are not aware of the first, and only the second involves visible and willed changes in the external world: and it is here, in the transition from volition to macroscopic action, where the trigger principle proves helpful in explaining something of the *outward* path, at least, of the psychophysical dynamics. (The opposite path, as great a riddle, was not further discussed.)

3. *Aspects of Quantum Theory*

Passing over to quantum mechanics, particular emphasis must be put on this tenet of the theory: it is not possible to know the state of a physical system to such a degree of completeness that its future states can be univocally predicted. The barrier is fundamental and not in the nature of a temporary halt in the perfection of observational techniques. It is indeed an integral part of the theory itself.[5] Familiar expressions like "state description," "determined," "prediction," assume an altered meaning. There can indeed be a description (called the Schrödinger function ψ) of the state of the system such that, if it is known for a particular time t_0, it is also known for any future $t' > t_0$—but only if no new measurements are made. Even if new measurements are made, probabilities for their outcome are determined by the Schrödinger function obtained from the original measurement. This sounds "classical" enough. However, the information thus provided is not what classical physics would have called a complete description of the state of the system for all times $t' > t_0$ if the state at t_0 is known. It merely allows us to compute probabilities for the outcome of certain measurements (or of certain interactions of the system with other systems) in the future. This is what "state of the system at t_0" *means* in the context of quantum theory. In some cases, the possible outcomes of measurement are very sharply defined, and thus we can even say that they will assume, with great precision, only particular values, for example, the future energy state of an atom will predictably be one of the values $A, B, C, \ldots$, but none in between. Among the eligibles themselves the choice for coming out a winner is open and only subject to probability gradients. Thus prediction, derived from ψ, of the outcome of a measurement at $t' > t_0$ is *not univocally* defined. If we still wish to speak here of causality, we must note that this concept differs now from causality in classical mechanics in two striking respects. First, it is on principle impossible to measure simultaneously the values of all the quantities connected with the system, such as position, velocity, energy, etc., and no set of future measurements can recapture all of the values of the other parameters at the moment of the first measurement.[6] Second, in general, any sharp measurement changes the "state" (in the sense defined) of the system. What one can measure—to any degree of exactness—each time is *one* of certain complete sets of compatible quantities that are said to determine the "state" or the Schrödinger function of the system at the given moment. ("Compatible" means: simultaneously measurable without mutual interference.) Then the "state" defined in this narrow sense is indeed determined in the future—but if, at any later time, one measures one of the incompatible quantities, the state "collapses" and is replaced by a new state description. Also, the sharper the values of one set of compatible quantities are measured, the less sharply known are the simultaneous values of the alternative set: from a middle ground of equal inexactitude for both, the ratios of simultaneous knowledge for the incompatible quantities stretch to the opposite extremes of all—zero, where the product of the indeterminacies is constant (Heisenberg). In sum, at no one time is the state of the system, and therewith the outcome of future measurements, completely knowable; it is, in this sense, an "ultraphysical"[7] reality (Friedrichs).

4. *Possible Uses of Quantum Theory for the Psychophysical Problem*

The bearing of all this on our subject is that here, at a crucial spot in the texture of things, a gap parts the web of our knowledge, a gap not given to being closed on principle, but sufficiently defining the *terra in-*

cognita within for us to understand that in it we can no longer appeal to a deterministic physical theory.

The attempt to locate *in this "gap"* a solution to the psychophysical problem, or to that of "freedom and necessity" (or there to seek an *asylum ignorantiae* for it), is not new.[8] The tendency appeared almost from the beginnings of quantum mechanics and principally fastened (as far as I am aware) on two aspects of the theory: on the principle of complementarity and/or on the principle of indeterminacy.

a) Complementarity   Niels Bohr himself, who in his philosophical moods was much inclined to extend the formalism of his complementarity principle to areas beyond its native soil, at one time tentatively suggested its possible application to the psychophysical problem,[9] never (to my knowledge) to take it up again. But the wide currency which "complementarity" has gained, not without Bohr's encouragement, as a generalized logical tool for "dual aspect" situations of all kinds outside its home ground (notoriously so in the social sciences), places it in the field as a candidate for tackling the primordial "dual aspect" case of human experience, the psychophysical. I regard all these extramural uses of "complementarity" as highly dubious and at best metaphorical; for the psychophysical problem I will try to show that it logically does not fit the situation at all.

First a few words about the original, quantum-mechanical sense of "complementarity" as coined by Bohr. It concerned the possibility of defining the "state" of a system in terms of two, mutually exclusive, conceptual representations. One may say that complementarity, in the sense of Bohr, means that we can ask one of two mutually exclusive sets of questions about a system, but not both. The answer to one question would describe the system as a particle; the answer to the other would describe it as a wave. But the two models are not just optional alternatives equivalent with one another. They stand for different observables. For example, the position component X_1 is complementary to the impulse component V_1, thus the particle description answering to the measure of X_1 is complementary to, but not interchangeable with, the wave description answering to the measurement of V_1. The knowledge of one of them precludes the full determination of the other, and the more I know about the one, the less I know about the other; yet both are required for a full account of the phenomenon, complementing one another in conveying its truth, that is, the exhaustive knowledge of what is knowable about it. As quantum physicists are wont to say: The entity "is" a particle when I measure its position, and it "is" a wave when I measure its momentum. Thus, whatever it is "in itself," only a dual account can do justice to the object (or, express the "truth" about it), without therefore bespeaking a dual nature of things.

Now it is tempting to think that something similar might apply also to the twofold account of human action (and of conscious behavior in general), the "outer" and the "inner," and provide a solution to the ancient problem of necessity and freedom: descriptions of one and the same train of events in terms of physical necessity, on the one hand, and in terms of mental spontaneity, on the other, are "complementary" in the sense of Bohr's principle; the either-or is one of representation, not of fact, and only both representations together *in* their difference convey the truth of the identical fact. As in the case of the particle and wave descriptions, both are equally genuine—and, we should add, equally symbolic, unless we are prepared by analogy to apply to this situation too the boldly idealistic assertion of certain quantum physicists that the "object" itself actually changes, with their measurings of it, from "being" a particle to "being" a wave, or vice versa (which I doubt even they would mean to hold for the corresponding terms in the mind-body "complementarity").[10] Somehow, at any rate, the underlying reality thus doubly expressed is supposed to be one in itself.

To discuss this appealing suggestion thoroughly is not appropriate here. It would turn out to "work" as little as any variety of "psychophysical parallelism," Spinozist or other. One little observation, however, may suffice to show that the transfer is in this case logically at fault from the outset, namely, not faithful to the formalism of the original concept. To "complementarity" in its quantum physical conception by Bohr himself, it is essential that the two descriptions are cleanly separate, each complete in itself and neither intruding into the other: the wave description is not to be contaminated by corpuscular terms, and vice versa. The two models are, in short, strictly alternative. But we cannot begin to describe anything "mental" without referring to the "physical," the world of objects with which mind, sense, will, action have actively and passively to do. That is to say, any speech about mind *must* also speak of body and matter. And when speaking of ourselves, we not only can and do but always *must* embrace with it our physical and mental being *at once:* precisely that *simultaneous* entertaining of *both* sides which quantum theory rules out for its complementary alternatives. From this original, *joint* givenness, after all, the psychophysical problem arises in the first place. Here, the *isolation* of the two components is an artifact of abstraction, their interlocking copresence being the primary datum. Even in abstraction, we noted, the isolation does not really succeed, as the description of one side intrinsically refers to the other. The lines themselves do not run parallel but cross.[11] On this *transitive* relatedness alone, by which one description draws into itself elements of the other, the purported analogy with the quantum-mechanical situation breaks down.[12]

More to the point of our discourse than this formal observation is the blunt reminder that what is substantively at issue in the psychophysical problem is *interaction* and, more particularly (surely so for Bohr's interest in "freedom"), the question of an *intervention* of mind in the affairs of matter. This, if it takes place at all (which is just in question), is nothing like an invariable concomitant, an innocent complement of physical processes, but is a particular event affecting their course. Does this happen? Is it possible? How? Such a question is evidently meaningless in the case of the complementary wave and particle description—for example, to ask whether, to what extent, and on what occasions the wave aspect of events leaves its mark on their particle aspect. But just those questions (with the appropriately substituted terms) are the most meaningful ones to be asked in the psychophysical setting.[13] Complementarity, "noninterventionist" by its formal nature, does not even allow them to be asked when seriously held to apply. In sum, I believe, no faithful analogue of complementarity as understood by Niels Bohr really applies to our problem, and philosophers should leave it where it belongs. The philosophical interest of its attempted enlistment lies in what it has in common with Spinoza's parallelism of attributes, namely, the *noninteractional* premise, of which it seems to offer a more sophisticated version. But it shares with it the fatal weakness of every parallelism: to leave the body in the dominant position, all assurances to the contrary notwithstanding. For only the body's internal determinism, that is, that of material nature in general, is *known* with the force of predictability, while that of the mind is not. By the logic of theory as such, the "hard" side thus distinguished will always prove the stronger partner in a marriage with a nondeterminist, "soft" mate, who can only play second fiddle to the other's leading tune. Matter will dominate and mind has to follow suit. This is the demonstrable fate of every form of psychophysical parallelism, of which "complementarity" is a variant.[14] "Epiphenomenalism" lurks in all of them. Thus, quite apart from the formal defects, complementarity, in the way it materially predecides the crucial issue, has taken the sting out of the problem that should not let us come to rest. Or, more likely, it is nonrigorously applied, and in that vagueness it is worthless.

b) Indeterminacy　　A better promise is held out by the principle of *indeterminacy*. To enlist it for the theoretical cause of "freedom," seeing that this is somehow ranged against "determinism," is an obvious choice, and it appeared early in the literary wake of quantum mechanics. Almost as early, however, two grave objections were raised against the idea. First, the working of mind—thinking, making decisions, etc.—even if not determined by physical necessity, is anything but indeterminate; its freedom goes together with high-grade orderliness, and so no principle of randomness can profit a theory of that freedom.

Second, no contribution from that quarter could make itself felt on the plane of macroevents where human acting takes place: there, the latitude of probabilities obtaining on the subatomic plane, where quantum me-

chanics governs, is superseded by the tight determinism of classical mechanics, as with the statistics of great numbers the probabilities from the
microsphere turn into certainties in the macrosphere. So, in that respect
too, the indeterminacy principle offers no comfort to freedom in that
sphere, and none other is worth talking about.

The first of these objections, that of randomness versus the orderliness
of mentality, can only be countered by hypothesizing that mind has somehow the power to bend indeterminacy to its purpose—to pick out, as it
were, the winner from among the spectrum of probabilities. How to represent, by a model conception, the manner of such an intervention or
influence is a matter for speculation and perhaps for despair; but to try
for it in the open spaces outside verifiability is surely not interdicted by
any veto of natural science or of rational thought in general. Even failing
there (as may be fated), the hypothesis as such would not *clash* with the
data and rules of that physical locus and may as well enjoy the licence
of the *terra incognita* (incurring its dangers too). To this, we shall return
at the end.

The second objection, that of the statistical submersion of subatomic
indeterminacy, can be more positively met if a way can be shown on
which a single quantum-mechanical event may become a determinant for
events on the macroscale of our experience and action. Now the *trigger*
principle offers just such a way, and herewith I resume the digest of the
sessions with Professor Friedrichs.

5. *Quantum-Mechanical Hypothesis about the Brain, and*
the Idea of Replicating One

What would follow if we entertain (for argument's sake) the hypothesis
that the brain is so organized that for its working—and then for the behavior of the visible organism—transactions on the quantum-mechanical
(subatomic) level can become relevant? Even before considering *how* they
can do so, one surprising inference was pointed out by Friedrichs. I had
spoken (above p. 221) of "a brain of the same physical constitution as
the human, but without concomitant subjectivity," and then discarded the
very hypothesis of a "merely physical brain," my point there being the
nondissociability of a psychic complement from a living brain. But exception should be taken already to speaking of a "same" physical constitution as . . . with reference to the brain at all, which recalls the Cartesian
hypothesis of exact "replication" treated earlier in this Appendix. For
exact replication presupposes exact knowledge of all constituents of the
system to be replicated. But if, *ex hypothesi,* such a system is really
defined, in the causally meaningful sense, by quantum-mechanical state
descriptions, then the exact, that is, complete, knowledge is on principle
beyond reach and the very idea of an exact replication is inadmissible. It
is—in the strict sense, and not just as a toll of excessive complexity—

impossible to have so exact a knowledge of a state of a human brain that we could predict all its future performances. Thus, also, one cannot design it, because in order to design it one must know it. But for what one *can* design there holds then the corollary truth that it will inevitably be a deterministic system. (This would not be basically altered by building in a random factor.) To quote Friedrichs: Our knowledge of the physical state of the human brain *can* only go so far that, in imitation of it, we construct a robot. In sum, if "state of the brain" means indeed "quantum mechanical state," then the famous Cartesian thought test of a perfect physical replication (by none less than the Creator, the perfect mechanic himself!) breaks down on this object. Thus, even without considering the role of the "subjective" factor (my original argument), robot would in terms of physical function alone, that is, in "overt behavior," always remain robot.

6. *Indeterminacy, Trigger Chain, and Macrobehavior* ("*Schrödinger's Cat*")

But how can "decisions" on the quantum-mechanical level determine macrobehavior? My model employed the trigger principle, and this was somewhat further elaborated by Friedrichs. First, the principle can be applied repeatedly and serially: triggering of trigger of trigger . . . , beginning with an arbitrarily small amount of energy. This I had in mind myself when speaking of the organism as an "amplifier." (Already Whitehead had done so.) But as long as classical physics is given unqualified validity, the trigger series will not lead back to something new: even its near-zero origin will still be subject to the general deterministic laws. However, "Schrödinger's cat" shows that with the trigger principle we can easily pass into regions where quantum physics takes over.

What is known as the problem of "Schrödinger's cat" was devised by him to illustrate the difficult problem of the role of the observer in quantum mechanics.[15] We shall use it here for a slightly different purpose, as it also illustrates, in a down-to-earth example, the difference of "causality" in classical and in quantum mechanics and at the same time shows a way in which the latter can be made to intrude into the domain of the former. The situation imagined by Schrödinger is the following. In a box, there is a cat, a vial of prussic acid, a sample of radioactive material, and a trigger mechanism that will break the vial (and thereby kill the cat) if, say, an alpha particle emitted by the radioactive substance hits a certain disk at the starting end of the trigger chain. If we know something of the state of the radiating material at a given moment, we may be able to compute with great accuracy the chances of the cat still being alive after one hour. Let us say that the probability is $1 \div 3$. What then can we say of the state of the cat after one hour, on the basis of our initial knowledge of the system at t_0 (i.e., ruling out our looking at t_1 through a window in the box)? Only that the cat "is" one-third dead and two-thirds alive.[16] In classical mechanics we would have been able to predict the exact moment of the cat's death. But in quantum mechanics we cannot predict *when* "Schrödinger's cat" will be killed, nor later reconstruct when it was killed.

Now, of course, if there had been 1,000 boxes, with their initial conditions as nearly identical as they can be made, the pronouncement after one hour would not have been that absurd one (which it is with respect to one cat), but the entirely nonparadoxical one that two-thirds of the number are alive and one-third dead; and upon inspection, this would be found nearly true, the more nearly so the more boxes there are. Thus the quantum mechanically induced unpredictability of the single case vanishes again with increasing numbers. But in the question of "freedom and necessity" (or "mind-body" in general), not populations and averages matter,[17] but precisely the individual, just as with regard to the single cat it matters whether *it* is alive or dead, and *when* the latter contingency in the yes-no alternative takes place. The meaning of Schrödinger's thought experiment, as here used, was precisely to be a single-case experiment. Thus it illustrates what we are concerned with: "indeterminacy" carried over—by high subtlety of organization—from the micro- to the macroorder.[18] *If*, then, as the hypothesis has it, the human brain *is* such an organization, it may enjoy, for the macrodetermination of the body, that is, of our behavior (as well as for the internal determination of its nontransitive activities in mere thought) whatever latitude the quantum-mechanical indeterminacy of its base level offers it to play upon. This, to be sure, as Professor Friedrichs took care to stress, does not *explain* action of mind on matter or interaction between the two (there is, in his words, "no theory of that" in all this); but it does remove the standard objection that this whole notion is unacceptable to physical theory and the occurrence therefore to be denied. In other words, it disposes of the "incompatibility argument" in the psychophysical problem and thereby of the exclusionary dictate of materialism. The gain, even if lying in the negative, is philosophically significant: in quantum physics there is no flagrant contradiction between mechanics and the influence of consciousness.

This is how far the actual discourse, of which this is an expanded protocol, got by way of "results." If nothing else, they leave the feeling, or a persuasive conjecture, that it must be "here," in the *terra incognita* of the quantum-mechanical dimension, where the mysterious switch takes place—from mind to matter and from matter to mind (in its two-way interchange role an odd reincarnation of Descartes' "pineal gland" of unhappy memory). Moving beyond this *raisonnement* of possibility and the demarcation of its locus to an "explanation" or, rather, representation of the transaction itself would need a theoretical model whose terms are not borrowed from one side or the other—a *tertium quid*, neutral to the distinction of matter and mind, prejudging neither one in the image of the other, but able to account for a transmutation, conversion—or whatever be the dynamical mode of transition—among the two. No such model is at present in sight. My all too crude metaphor of the osmotic "wall" could obviously not appeal to Friedrichs (nor does it much to myself). Something better is surely attainable, but a real theoretical solution may be destined to elude us.[19] It certainly did on this occasion. On this note of partial success and partial defeat, the discussion ended. As Friedrichs put it dramatically: "Having reached the point where I feel the solution must be looked for, I thought and thought—and finally threw up my hands."

Here, then du Bois-Reymond's *ignoramus* is presently lodged, perhaps even the *ignorabimus;* but better lodged than before, because freed of a spurious logical straightjacket. For myself (going back to what I said in n. 4), I add that even an *ignorabimus* in the terms of science need not halt in this matter the conceptual effort of speculative philosophy.

Notes

Chapter 1

1. Immanuel Kant, *Groundwork of the Metaphysic of Morals,* preface.
2. Ibid., chap. 1.
3. Ibid. (I have followed H. J. Paton's translation with some changes.)
4. Except in self-cultivation and in education. E.g., the practice of virtue is also a "learning" of its discipline and as such progressive; it strengthens the moral powers and makes their exercise habitual (as the converse is true of bad habits). But naked primal nature can always break through again. The most virtuous can be caught in the destructive tempest of passion, and the most wicked may experience conversion. Is the same still possible with the cumulative changes in the conditions of existence which technology deposits on its path?
5. On this last point, the biblical God changed his mind to an all-encompassing "yes" after the Flood.

Chapter 2

1. As far as I can see, this has been paid too little attention in moral philosophy. In the search for a concept of the good, with which moral philosophy is concerned, it has tended to consult our desiring (with the Socratic premise that what is most desired is precisely what is best), while actually our fearing would have been a better guide. The "eros" of Plato, the *appetitus* of Augustine, which by nature seek a good and ultimately *the* good, are examples of the appeal to desiring. That may be correct in the end for a fully informed desiring that also fully knows itself. But how do we come to know our desiring? By noting the desires as they occur? Certainly not. Which of these two I desire more: that my daily meal taste good or that my child remain well, that I cannot at all gather from the felt and compared strengths of the desire in either case (one of which makes itself felt daily and the other need not do so at all). But if I must *fear* for the health of my child, because I have suddenly reason to, then I know.
2. It is nothing other than a version of the Cartesian argument of an evil and imperfect creator of our existence (for whom, Descartes himself said, one can substitute a blind and value-indifferent Nature). Its prototype in turn is the ancient argument concerning the Cretan who declares all Cretans liars.
3. The complete argument would continue: and unless, within being, something better

than man would take his place upon (and conditional upon) his disappearance. But the step is redundant if already the primary case for "being as such" entails (as it will later be shown to do) that man represents the maximal realization we know of within given reality.

Chapter 3

1. Technological intelligence thus can go both ways, from end to means and from means to end, and answer the question, "How must a thing look if it is to fulfill this and this purpose (e.g., drive in nails, measure time)?" as well as the converse one, "Which purpose can a thing have that looks so and so?" It is precisely such a purely objective, as yet neutral "look" which is lacking in the other class.

2. Hence the question, "What is a hammer?" can be answered with a picture (as is done in primers and dictionaries), but not the question, "What is a legislature?" In its case, we can't even explain its name without divulging its purpose.

3. If Aristotle in one place (of contested authenticity) in *De anima* goes so far as to designate the entire body as a "tool of the soul," then this is already a dubious transfer and actually not in good agreement with his prevalent biological use of the "tool" concept, in which the *parts* of the living body are tools of the whole, that is, of just this ensouled body. Besides, it also contradicts the Aristotelian concept of soul, as the *immanent* "entelechy" of the body, to describe its relation to it in terms of the user of a tool, which is to make it an extraneous relation.

4. With this interpretation, of course, also the teleology first admitted to operate in the *making* of the machine sinks back into the realm of mere appearance: the makers in turn are machine-building machines.

5. Here, Descartes was tactically wise but left with a foolish substantive position he had to adopt in consequence, when he limited his dualism—and hence even the most elementary presence of feeling—to the mind-body relationship in *men:* the consequence for all the rest of the animal kingdom was the "beast-machine" with no subjectivity at all.

6. The "holism" of Jan Smuts should also be mentioned here.

7. The principal orders of evolution, e.g., cannot be "predicted" (deduced) from the beginnings even retrospectively, i.e., with a knowledge of the actual results. Of a deductive sufficiency can here, so far, not even be a pretence. Strictly speaking, we do not even know *why* atoms for the first time formed the double helix of the DNA molecule; only the fact of their having done so tells us *post hoc* about the possibility thereto and with this teaches us something about the laws of molecular structure.

8. Leibniz already saw all of this, though in the dyad of *perceptio* and *appetitus* he gave the chief significance to the first.

9. Here the paradigm of junctures with "causal indifference," presented in the Appendix in the context of the psychophysical problem, can perhaps be of service too. Accordingly, evolution also might be understood as a series in which critical equilibrium thresholds occurred by the thousand and, with their several, causally equivalent alternatives, enabled a hidden tendency to exercise its "preference" and resolve the momentary indifference in favor of one of the open options. This would then be the meaning of the term "opportunity."

10. H. Jonas, *The Phenomenon of Life: Toward a Philosophical Biology* (New York: Harper & Row, 1966; Chicago: University of Chicago Press, Midway Reprint, 1982), Third Essay, especially pp. 80–86.

11. We said in advance (p. 65) that a lesser certainty of answer was to be expected here than to the question of the power of purpose in voluntary action. The latter we could *prove* by the conclusive refutation of the opposite assumption (see Appendix). Here we could only plead the eminent reasonableness of the assumption as compared to its negation.

Chapter 4

1. This is somewhat of an *argumentum ad hominem,* insofar as it exploits a spontaneous preference for one of two logically possible alternatives. But perhaps with this very bias it restores the balance to a matter which, in the preemptive course that philosophical theory has taken in its long seclusion with natural science and itself, has not had its proper say for most of the time.

2. That the living being is its own end does not yet mean that it can *set* itself ends: it *has* them, by nature, in the service of the unchosen end-in-itself. Any simultaneous serving of the ends of other beings, even of its own brood, is only coincidentally and by way of genetic disposition included in the pursuit of its own: the vital ends are selfish from the viewpoint of the subject. (The objective subordination of these individual ends under more comprehensive ends of the biosystem, their fitting into it, is a separate matter.) Only human freedom permits the setting and choosing of ends and thereby the willing inclusion of the ends of others in one's immediate own, to the point of fully and devotedly making them his own.

3. *"Die Wanderratten"* by Heinrich Heine, which begins with the lines, "There are two kinds of rat, / The hungry and the fat."

4. Ends are not necessarily chosen, and still less so by comparative evaluation: acting as such (animal behavior included) is guided by ends even prior to all choice, since elementary ends—and having ends at all—are implanted in us through the neediness of our nature. And the accompaniment of pleasure makes them subjectively "valuable" as well. But the present discussion is about the human sphere of *chosen* ends, where willing is not simply a creature of the given end, but rather the end—as my own—is a creature of willing. Even here, the "value" of the end is a correlate of desire, which is variously predetermined itself—by instinctual drives, environment, example, habit, opinion, and the moment.

5. That man's will is responsive to ends beyond his own vital ones—a marvel distinct from, but connected with, the natural marvel of reason—makes him a moral being. This responsiveness supplements and delimits the indifferent freedom of reason. As pure intellect, that is, as will-free cognitive faculty, reason can contemplate the world from the distance of neutral knowledge without taking a position; as technical understanding, it can devise the appropriate means for whatever ends the will seizes upon; but as faculty of judgment, instructed by sentiment, reason weighs the possible ends according to their worthiness and prescribes them to the will. Ultimately, however, the will already stands behind all these forms of reason. It is the will to objectivity which makes possible the so-called neutral cognition; and the will to ends in general, first to one's very own, which bids technical understanding to seek out the means; and the will to worthy ends, which bids the faculty of judgment to listen to feeling. Of the will in this primal sense, it is perhaps true what Nietzsche said of it: that it would sooner will nothingness than desist from willing. But for being able to will *something,* the will (or the "judgment," to which it is ready to listen) needs the sentiment, which bathes this something in the light of the choiceworthy.

6. Despite nominal similarity, Max Weber's distinction between ethic of responsibility and ethic of intention does not fall within the above dichotomy between object-ethic and subject-ethic. For what he describes as the "ethic of intention" (*Gesinnungsethik*) and contrasts in politics with the "ethic of responsibility" is merely that single-mindedness in the pursuit of an *object* (a "cause"), posited as absolute, which heeds no consequences except the hoped-for success, and to which no price (payable by the community) seems too high, and even the risk of failure with its total debacle warrants the attempt. The "politics of responsibility," on the other hand, weighs the consequences, the costs, and the odds, and never says of *any* goal "may the world perish, but let justice be done" (or whatever else the highest good might be). But he who does say it is, naturally, dedicated to an *object,* and—since he considers it realizable—he has the common good (as he sees it) no less in mind than his more moderate counterpart. As a matter of fact, the Spartacists, of whom Weber was thinking at the time, saw themselves as thorough realists; and to Rosa Luxemburg

it was neither purity of purpose nor fidelity to the program which mattered, but the seizing of a chance, great or small, which to pass up would have been in her eyes a betrayal of the greatest of objective causes. That she paid for it with her life does not make her undertaking irresponsible (though perhaps unjudicious). Hence it is merely the distinction between radical and moderate politician, between one who knows only *one* goal and one who tries to strike a balance among several, or between one who stakes everything on one throw and one who spreads the risks, which Weber expressed in the famous pair of concepts, "ethics of intention/ethics of responsibility." (The fact remains that one-sidedness and fanaticism are unfavorable conditions for responsibility, which requires circumspect judgment.) The distinction is important enough and will concern us further *inside* the ethic of responsibility (especially in connection with the ideal of utopia), but it falls as a whole within *one* side of the dichotomy which we presented above. What Max Weber had indeed to contribute to the problem of ethical *subjectivism,* which likewise lies within that side (i.e., that of commitment to objects) but cuts across *his* dichotomy, is his thesis of "value-free science" and its "demagified world." In fact, behind the nihilism of existentialism and its ethic of artbitrary value-setting, just as behind the whole of modern subjectivism, stands modern natural science with its premise of a value-free world.

7. For a powerful critique, see Max Scheler, *Der Formalismus in der Ethik und die materiale Wertethik* (Halle, 1916).

8. The same word for two so different meanings is no mere equivocation. Their logical connection is that the substantive meaning anticipates the full force of the formal meaning to fall on the agent in the future for what he did or failed to do under the substantive mandate.

9. Shared danger surely establishes *reciprocal duties* of their own kind. But as long as I was not unilaterally the cause of this danger or of a particular peril in the course of the enterprise (and have thus become "responsible" for *this* in even the mere formal sense), those duties are in general those of a situation in which everyone must be able to count upon the others. To fail here out of weakness is a sin against loyalty and the other virtues which the trial of the situation may require (as courage, resolution, constancy) but not strictly against responsibility. Strictly "irresponsibly" I do act when I endanger my comrades and the whole enterprise by an act of positive folly—which then also makes me causally superior to all the others, on whom I have inflicted my deed unilaterally.

10. The "public," i.e., political, office (of representative, cabinet minister, president) must be distinguished from the technical office of the functionary (civil servant). It is the distinction between government and administration, and it is the former which is the subject of the electoral process. The right to vote can, if you will, be understood to carry with it the duty to exercise it; the right to run for office merely denotes the formal qualification for candidacy, but not a duty thereto. He who stands for election does so by his own choice and has first of all elected himself. To be sure, Athenian democracy even made it a duty for those chosen by annual lot to accept the office. But this was before the distinction between political and administrative function of magistrates: *every* office was "political," i.e., given the assumed equal fitness of all citizens for the affairs of the city, any public office was considered within the competence of any of them. But the real power in determining the *politics* of the commonwealth lay with the "demagogues" (in the prepejorative sense of the word: popular leaders) who, while obtaining their mandate from the electorate, were in the final analysis self-chosen.

11. The decision is morally self-evident because the work of art cannot ask me the question, "What have you done with me?" while the child, in my imagination, can ask it of me (e.g., as accuser before God's court), and him I owe an answer. In the case of the work of art I have only to face the deprived art-lovers, and them I *can* answer, namely thus: this was the situation, and henceforth, regrettably, you will have to be content with reproductions. But I cannot answer the child, "You will have to be content without your life," and so I

cannot answer him at all and thus cannot defend my choice. Over against this incommensurability of rights, the reflection that the child might possibly grow up to be a good-for-nothing who was not worth the sacrifice of the Sistine Madonna, is as irrelevant as the thought that he could possibly become a genius, perhaps a greater one than Raphael. Absolutely nothing follows from either, nor from the relative probabilities (good-for-nothings are more frequent than geniuses). I mention this weird thought-game because it plays a role in the literature (see Arnold Brecht, *Political Theory,* p. 154, where Radbruch is quoted, who quoted Sir George Birdwood, who for his part—under the protection of the hypothetical—decided in favor of the painting), and because recently, in conversation, an esteemed scholar of tried moral integrity showed himself seriously torn between the alternatives and at least tempted to side with Sir George.

In reality this intellectual exercise demonstrates that the concept of value cannot form the sole basis for a doctrine of duties. The life of the child is neither for him, nor in itself, nor for others, a "value" as the artwork is for its beholders and may perhaps once be so for the child as well. If, as is the custom now in the Western world, one speaks of the "infinite value of each human life" (an afterglow of what the Christian religion said of the soul in God's sight), then this can sensibly only mean the *right* of every life to itself, its right to the end-in-itself which it is. That right, to be sure, is not "infinite," seeing that its object is finite, but it is "unconditional" in the sense that it is (*a*) derived from no other source, (*b*) independent of qualifications (including "value"), and (*c*) that no one else has a right to the same thing. There can be a right to the value of the artwork, e.g., that of all its potential enjoyers, but none to the life of the child except his own right thereto (though there *are* rights of others to the manner of conduct of this life). This truth is not contradicted by the fact that superior claims—of public peril, bearing witness to the faith, even of a code of honor—can demand of the individual to set aside his primary right for theirs (as in war), nor by the fact that it might be forfeited through deadly crimes. Rather it is part of one's own primary right, precisely insofar as it claims recognition, that it includes the recognition of other rights and therewith the acceptance of duties, even unto readiness to die. All this has nothing to do with value-theory, although values naturally can become duties.

12. This is a modern development of the secular state, but before this the church—likewise a public institution—performed approximately the same task.

13. The case of Moses is the reverse: out of community of descent he desired for himself solidarity of fate with his enslaved kinsfolk.

14. Not meant here is religious eschatology (such as Jewish messianism), which precisely is *not* "immanent" theory of history.

15. "Every epoch is immediate to God": Ranke against Hegel.

16. The biblical simile for this is Adam, *created* in the image of God—certainly a presence on earth and not an expectation for the future, whatever in the way of unearthly transfiguration might be expected from the "new Adam" of the last days.

17. Personally, I would guess: no.

18. First represented by Hegel and Comte at about the same time.

19. In natural science this is otherwise. If all past experience, including that controlled by experiments, has proven a certain regularity, then this is also proven for the future. For nature, so we assume with good reason, does not change (without its nonwhimsical uniformity, there would be no natural science) and, besides this, is not influenced by my opinion of it. With regard to history, however, uniformity is an (at least) problematic assumption, and its being influenced by the opinions of the historical subjects themselves, hence by their theories about it, belongs to its own "causality" (see further below).

20. Only one simplistic hint: no demonstration that the world must exist is confirmed by the fact that it does exist.

21. Other discrepancies with the theory: the idea of a "socialist fatherland," and the (still

incalculable) fact that a conflict with the communist brother, China, could become more probable than one with the capitalist opponent, America.

22. Of the theories of history with predictive claims, the Marxist is the only one which has *practical* implications, the only one therefore which must be considered in the thematic context of political responsibility. How theories of history without this element of self-fulfilling activity fare with their predictions can be seen in the example of Oswald Spengler, whose ambition it was to elevate historical science from mere knowledge of the past to prediction of the future, and who believed to have attained this through his organic-mor-phological method. Here the biological "life-stage" simile is at home and the future is as determined as individual aging is, of which nothing can be changed either, whether one knows of it beforehand or not. Logically Spengler was right in holding that only a fatalistic pattern of history allows predicting the future. But what of his predictions? One as general as the rise of Russia was already professed, as a surmise, during the nineteenth century by people with and without theories of history (or, as a promise, by Pan-Slavist prophets such as Dostoevski): something which for the discerning eye loomed on the horizon, for which certain conditions already visibly existed, and which by the short yardstick of current events has meanwhile indeed come true. Spengler's principal prediction, peculiar to himself, of a Russian world millennium replacing the declining West must await its time. All that can be said today is that in the world of only half a century later, with the further "Westernization" of the world—*including* Russia—in full stride, the whole idea has a fossil look about it. Still, the story is not yet told. But where, rather than millennial vistas, predictions that truly flow from the theory and nothing else can be tested over the short run, the results are plainly embarrassing. Thus the assertion, dictated by the theory, that Western mathematics had exhausted its possibilities and nothing more could be expected of it, came at the very moment when one of the most creative developments of this mathematics, opening up entirely new horizons, was just beginning. It is similar with the alleged dropping-out of history, the definitive "fellahization," of entire, once historical populations—a doctrine with which the otherwise soberer epigone Arnold Toynbee allowed himself to be infected. Not to speak of the mighty example of China, he cannot forgive the tiny Jews for daring, in violation of his theory, to again become an active subject of history and, after the accomplished deed, exerts himself on the thankless proof that what must not be, cannot be. (Spengler, by the way, following Nietzsche's example, had a more open mind concerning the Jews.) Fortunately, Ben Gurion paid as little heed to Toynbee as the mathematicians of the twentieth century did to Spengler.

In the light of the perpetually surprising course of events whose witnesses we presently living (often surviving) have been, one would then also regard the *ex post facto* part of the historical constructions, which proves that everything occurred as it had to occur, as a pastime of reason—without denying that sometimes, in intelligent hands, it can contribute something to an understanding of the past. To what extent Marxism here represents a special case was shown above.

23. Strictly speaking, of course, this holds for the education of children too, but there as we indicated—with the perennial "new beginning" afforded by the resources of personal spontaneity—antecedent deed has not the same finality of "results."

24. Naturally, dynamism does not *per se* belong to the collective human condition: its rule is itself a historical phenomenon, hence in principle also subject to the historical pos-sibility of again making room for another condition. In its present form, the phenomenon is without precedent and has much—in its exponential growth rate perhaps everything—to do with the eruption and "self-propulsion" of modern technology. This is not to say that earlier history flowed along evenly. But even its dramatic convulsions and crises, as those caused by major mass migrations, where things proceeded "dynamically" enough for a period, are not to be confused with the self-propagating dynamism of our age. It was much more coercion from without than internal impulse which drove them, and with migration

and conquest once completed everything gravitated toward earliest permanency. Spectacular changes in modes of existence, with the awareness of those undergoing them, were the exception; other vicissitudes were usually calamities of one side as the obverse to the luck of the winners. The long-term gradual transformations even of static civilizations, visible only to retrospection from a great distance, but unnoticed by the contemporaries, do not belong here, where we discuss the optical range of sociopolitical planning—the time range of one generation. And there, the crucial fact is that change *and* awareness of it, novelty occurring and furthermore expected, belong to *our* daily lives.

25. Even then it can fail to occur. There is perhaps nowhere a more goal-defined research program in existence, one more lavishly endowed with talent and means, than the present cancer research in America (to take a nonpolitical example). And yet, it could be that it will never reach its final goal because "the cure for cancer" is perhaps just not in the nature of the case.

26. To this also belongs the confidence that "technology" will always get the better of the problems it has created itself—that only a yet more advanced technology is needed in order to find the remedy for the evils with which past and present technology has presented us.

27. This *Weltanschauung* can still make a practical difference in the way in which the survival dictate in question is complied with—for example, how willingly or reluctantly one enters into the authoritarian order if such has been recognized as unavoidable.

28. A completely irresistible "ought" would no longer be this but a "must."

29. See my inquiry "Is God a Mathematician?" (in *The Phenomenon of Life*, Third Essay) for a detailed discussion of the methodological and epistemological questions.

30. This is as true of the "average person" as of the genius.

31. But it is a psychological fact that the greatest fear of the automobile driver is to run over a child.

32. The child cannot really ask the parents, reproachfully or otherwise, "Why have you brought *me* into the world?" (for they had no influence on the particularity of this "I"), but only "Why did you bring a child into the world?"; and to this the answer is that incurring this "guilt" was itself a duty—not, to be sure, toward the not yet existing child (there is no such duty), but toward the morally binding cause of humanity as a whole. We will treat of this later.

33. Whether it is this also in individual psychogenesis (as I suspect) is a question of fact, which can be empirically investigated only when the first sexless test-tube person comes forth and one can observe whether feelings of responsibility develop in him.

Chapter 5

1. For more about this development, see "The Practical Uses of Theory," in H. Jonas, *The Phenomenon of Life: Toward a Philosophical Biology* (Chicago: University of Chicago Press, Midway Reprint, 1982), Eighth Essay.

2. The formula "*S* is not yet *P*," "The subject is not yet the predicate," is Ernst Bloch's shortest logical expression of his philosophy in *Philosophische Grundfragen I. Zur Ontologie des Noch-nicht-seins* (Frankfurt, 1961), p. 18. Compare Adolph Lowe's discussion "*S* ist noch nicht *P*" in *Ernst Bloch zu ehren*, ed. by Siegfried Unseld (Frankfurt, 1965), pp. 135–143.

3. Already Plato's *Republic* is in the question of public veracity a good stiffener against liberal naiveté.

4. More precisely, the presumable interest was, on the one hand, to strengthen the "Third World" in numbers as prospective allies against the capitalist world and, on the other hand, to so heighten its internal pressures—and that means: its misery!—that the explosion, naturally turning against the rich, will be all the more inevitable: a pitiless calculation of power-

politics, which would then contain an error only if in the meantime one came to belong oneself to the rich of this earth—and to this no Marxism in power anywhere would object. But playing with fire, i.e., with a possible world conflagration, is implicit in the concept of "world revolution."

5. One could also name the divinely playful irresponsibility, the will to try everything; but that would lead into a sphere which is irrelevant to the present discourse concerned with "responsibility."

6. The concept of progress has here undergone a precise parallel to the concept of "evolution" (= development), which also originally referred to ontogeny and, borrowed thence, was finally as good as monopolized for phylogeny (cf. H. Jonas, *The Phenomenon of Life*, p. 42f.). He who today hears the words "theory of evolution" thinks of Darwinism, not of the individual's unfolding his "imprinted form" in stages of growth. Likewise, whoever hears "progress" thinks of society and history, not of personal life histories.

7. Cf. Friedrich Engels, *The Condition of the Working Classes in England*.

8. From *The Three Penny Opera:* "Be a good person! Sure, who wouldn't rather be? / Give to the poor your own: why not?, I ask. / When all are good, His kingdom is not far. / Who in His light would not with pleasure bask? / Be a good person? Sure, who wouldn't gladly be? / But sadly on this planet where we dwell / The means are scarce and men are crass. / To live in peace with all, who would not like that well? / But circumstances won't permit, alas. / Alas, it's true what he has said / The world is poor and man is bad" (my translation). One also thinks of Heine's "Migratory Rats" (see n. 3 to Chap. 4).

9. Compare this with the contemporaneous and kindred teaching of Adam Smith about the "hidden hand" in political economy, which causes the aggregate counterplay of all the individual egoisms in their unrestrained pursuit of profit to operate automatically, through the laws of the market, for the economic good of the whole. Still earlier is Mandeville's formula "private vices—public benefits."

Chapter 6

1. A certain exception must be made, in the Marxist case, for Ernst Bloch, the most eloquent theorist of utopianism, of whom this chapter will treat later.

2. I am thinking here primarily of the newly liberated colonial peoples (mainly in Africa). Neither South Africa nor Latin America fits this classification. The relationship between colonial masters that have become indigenous and the ethnoculturally differing subject population, where economic exploitation fuses with political oppression and racial segregation, represents a special case which even in the presence of a modern industry falls outside the pattern of nationally homogeneous Western class states. The "class struggle" there assumes peculiar features, for instance, that of first emancipating the "natives," who here correspond to the "proletariat." But in the calculation of world revolutionary potential they belong together with the postcolonial peoples of the underdeveloped world.

3. That the plumber or electrician earns more per work hour than the college professor for whom he makes a home repair is not uncommon in America. And many a welfare recipient outdoes him in certain habits of consumption.

4. How in the face of this notorious fact Ernst Bloch could say, "The worker in the capitalist class state has always had only the freedom to starve" (p. 1061 of Bloch in n. 6 below), is incomprehensible, and casts a saddening light on what an intellectual can deem just to say in his outrage over injustice. Besides, it is an insult to the really starving in noncapitalist lands, who would be only too eager for such comfortable "starvation." It is as if nothing had happened in Bloch's own well-known world since Engels wrote *The Condition of the Working Classes in England*.

5. One thinks of the English and American example, but even already of Germany under Bismarck and Wilhelm II.

6. Ernst Bloch, *Das Prinzip Hoffnung,* 2 vols. (Frankfurt, 1959) (henceforth quoted as P.H.), p. 1055; compare also pp. 925ff. The naiveté evinced here in matters of natural laws will occupy us further. The German original of my book was given the title *Das Prinzip Verantwortung* in conscious antithesis to Bloch's influential book.

7. Only *solar* energy in direct use (or via wind and water) avoids *all* the risks mentioned: those of exhaustion and of environmental pollution by combustion products obviously, but the heat risk as well, as it only transforms radiation incident on Earth anyway. However, by all realistic estimates, the exploitable part of it can never supply more than a fraction of the voracious needs of an energy-intensive technical civilization (as we know it) enveloping the globe.

8. Karl Marx, *Das Kapital,* vol. III, book 3 (in Karl Marx and Friedrich Engels, *Werke,* 30 vols. [Dietz Verlag, Berlin 1976], vol. 25, p. 828 in chap. 48, "The Trinitarian Formula"). There is more at the same place on freedom equated with minimal expenditure of strength in a shortened workday: "Beyond [this residue of necessity] there begins the development of human power acting as its own end, the true realm of freedom." Consistent with this, and without misgiving, Marx hailed future automation (remarkably foreseen by him), with its growing elimination of man from the process of production, as a road to freedom, which is coincident with free time.

9. Ibid., pp. 592 ff.

10. Karl Marx, *Critique of the Gotha Program* [*Kritik des Gothaer Programms*] (Berlin, 1946), p. 21.

11. Marx, *Das Kapital,* p. 593.

12. Ibid., p. 600.

13. See n. 2 to chap. 5.

14. "Balance of colossal forces": thus Herr Stein in Joseph Conrad's *Lord Jim* when displaying to his guest an artwork of *nature,* a rare and perfectly beautiful butterfly.

Appendix

1. "Brücke and I, we have bound ourselves by an oath . . .": so wrote du Bois-Reymond in a contemporary letter to Hallmann, quoted by W. W. Swoboda, "Ernst Brücke als Naturwissenschaftler," in Hans Brücke et al., *Ernst Wilhelm Brücke. Briefe an Emil du Bois-Reymond* (Publ. Archiv d. Univ. Graz, 8/1) (Graz, 1978), p. xxxiv. The writer speaks of Brücke and himself as "conspirators sworn to make prevail" the above truth and thereby transform physiology into an "exact science."

2. "Über die Grenzen des Naturerkennens," delivered at the forty-fifth Versammlung Deutscher Naturforscher und Ärzte in Leipzig, 1872; printed in E. du Bois-Reymond, *Reden* 1, pp. 441–473. The *ignorabimus* is there pronounced on *two* questions of natural knowledge: the psychophysical problem, and the intrinsic essence of matter and energy (as distinct from the laws of their action). Du Bois-Reymond thinks it possible that the two limits of our knowledge are at bottom the same, i.e., that if we comprehended the essence of matter and energy we would also understand how the substance underlying them will under certain conditions sense, desire, and think. "But [he says] it lies in the nature of things that on this point too we cannot obtain clarity, and all further talk about it remains idle" (p. 462). Thus, the last word of the essay, intellectually and typographically, is "*Ignorabimus.*" Against the protests of outraged scientific optimism that greeted his verdict at the time, we must still, a century later, attest to his philosophical insight in recognizing that these very questions are *transcendent to physics as such.* See n. 4 (below) for more on this question.

3. This view of the matter would, e.g., rule out telekinesis and other spiritistic macroeffects.

4. Trying for such a reformed ontology would, of course, be that kind of speculation of which du Bois-Rymond has said that no clarity (i.e., conclusive evidence) can be obtained on its subject and therefore all further talk, beyond indicating the mere possibility of a

common root for the divided record of things (and perhaps voicing a theoretical preference for its parsimony), must remain idle (see n. 2 above). But granted the first half of the contention, namely, that "knowledge" escapes us here, the second half concerning the idleness of the pursuit does not really follow, except by the terms of natural science and its defined criteria of verification. These do not exhaust the space of intelligibility and meaningful inquiry. "Speculative philosophy," says Whitehead, "is the endeavor to frame a coherent, logical, necessary system of general ideas in terms of which every element of our experience can be interpreted" (*Process and Reality,* part I, chap. 1, sec. 1). Reason must make that endeavor, even though foregoing in it the kind of verifiability which the positive sciences enjoy (in different degrees: history, e.g., far less than physics). After all, "epiphenomenalism" itself is a piece of "speculative philosophy," only a bad piece, because it fails at least two of Whitehead's tests: that of "coherence," and that of letting "every element of our experience" be interpreted in its terms. Usually, in the grand and forceful systems of speculation, one of these tests gives way to the other: most usually, in modern times, the second—fullness of interpretation of all phenomena—to the first, logical coherence, which since the seventeenth century has predominated in Western metaphysics with a certain surgical ruthlessness. At any rate, Whitehead's requirements, or such as his, do provide standards for judging a speculative scheme and save the enterprise from mere fancy. To interpret reality in the light of *all* the knowledge we have of it, transcending a mere summation thereof, is a need, a right, and a duty of reason, and is different from exploring its particular provinces. The "need" was acknowledged by none more movingly than by Kant, but he denied reason the right to follow its innate thirst because no *knowledge* to quench it lay on its path. Whitehead, instead, speaks of "interpreting," and this—not quite the same as "knowing"—is not only unavoidable (for we do it anyway and on whatever level, primitive or sophisticated, articulate or inarticulate), but it also remains meaningful in the face of nondefinitiveness and without the blessing of ascertained truth. To return to our special subject: the psychophysical problem is one of the pressures with which multiform reality nudges reason beyond the safety of the severally uniform sciences into the quest for transcendent unity that can only be proposed and never proven; and after Descartes it happened to be the major such pressure to which the great metaphysical systems responded with their different solutions, all of them tinged with the characteristic violence of thought to which I have referred. More heedful to the shadings of experience, yet bolder by the radical conceptual reframing of ontology to do them justice, is Whitehead's grand attempt in our century. Its persuasiveness, besides the power of its internal coherence, depends on the measure to which it passes the external test set by himself—the oldest in metaphysics: "saving the phenomena," i.e., whether all of them can be interpreted without loss of character in terms of the system. It is no belittling of Whitehead's achievement, but a call to go on, to say that *not every* significant element of our experience is so "saved" in his conceptual scheme. But it is an inspiring instance, the only one so far, of what I have in mind when, over against my feigned "speculative model" framed in conventional terms, I surmise a truer one "framed in ontological terms which would alter our very speech of 'matter' and 'mind.'"

5. With reference to this aspect, one well may think of Niels Bohr's utterance of 1952 as reported by Werner Heisenberg: "If one is not at first shocked by quantum theory, one cannot possibly have understood it." Bohr missed that "shock" in the response of a philosophers' meeting in Copenhagen—mostly of the positivistic persuasion—to a talk by him (see Werner Heisenberg, *Der Teil und das Ganze. Gespräche im Umkreis der Atomphysik* [Munich: R. Piper & Co., 1969], p. 280).

6. Ideally, the state of the system at a given moment consists of all possible "observables," i.e., of whatever one can measure: position, momentum ("velocity"), spin, etc. All observables together, known simultaneously for time t_0, constitute what we may call the "Laplace state," which represents the ideal of classical physics. With that knowledge, held

possible in principle, all future and past states are also known, namely, determined. Just this simultaneous knowledge is held on principle to be impossible by quantum theory, notwithstanding the knowability of each constituent by itself. The very actualizing of the knowledge in one direction forfeits that in another (see continuation in text).

7. "Ultraphysical" renders the German *ultraphysikalisch* = beyond the grasp of physics, an epistemic term, not (of course) *ultraphysisch* = beyond corporeal nature—an ontological term which would place the "reality" in question in a different realm of being, e.g., the mental (in which case it would be eminently knowable!). The English "physical" collapses these two different meanings into one equivocal term. The nearest analogy in the philosophical vocabulary to the intended meaning may be Kant's "noumenal," referring to "the thing in itself," whose formal *concept* belongs to the "intelligibles" (= formed by the intellect alone), but whose content we cannot obtain. "Ultra-" was chosen in deliberate preference to "trans-" because of the latter's strong connotation of a transcendent, qualitatively different *kind* of reality (like meta-physical), whereas "ultra" can also mean more of the same, exceeding it in its own kind, as in "ultraradical," "ultraconservative." (This semantic clarification was provoked by the doubts of one very attentive reader of the manuscript, Professor Adolf Lowe.)

8. With the remainder of this section I depart from the digest of the talks, where this retrospective theme did not come up.

9. See, e.g., Niels Bohr, *Atomic Theory and the Description of Nature* (Cambridge, 1961), pp. 24 and 100 ff.

10. For the particle-wave alternative the conflation of epistemic with ontologic meaning, audacious as it is, is at least arguable, as the "objects" in question are *entia rationis* (theoretical constructs) to begin with. E.g., the wave aspect of the quantum-mechanical event, though mathematically isomorphous with the description of a concrete, physical wave, denotes so highly abstract an "object" as a probability wave, whose reality status and existential independence from the conceptualizing observer are indeed debatable. Nothing like it applies to such concrete, content-saturated data of our primary experience as body and mind.

11. This goes for both sides, though it is more obvious for the mental side by which we exemplified it: a sense perception is *of* a physical object and also (we assume) caused *by* that object. The physical side seems better isolable: one *can* describe the physics of the eye without any reference to seeing, and so with every part and even the whole of the physical organism. But I doubt whether it makes sense to do so for long (e.g., to treat of brain processes and not mention their mental implication), and even whether it is possible to speak of the quality-stripped entities of physics in general without in the negation *implying* the perceptual qualities from which they were abstracted. Also we must say, after all, that the pigment on the canvas causes the color perception in us to which it is correlated: this *acting on* the mind is as surely a statement about the *physical* thing as the being affected by it is a statement about the mind. In short, the two "sets," whether the one or the other is thematized, keep crossing over (or "interact"), and this is just the crux of the matter.

12. Even simpler, in this formalistic vein, is the objection that in the quantum-mechanical case we begin with data of the same kind (space-time measurements) and end up with a duality of representation of our own devising to account for them, whereas in the psychophysical case we begin with a duality of cardinally different data, not of our making at all, and try for a theoretical unification of them. If in such a unification they are found to be "complementary" in some sense, then "complementarity" itself becomes the unitary representation for a dual phenomenon. Thus, the direction of the logical operation in the two cases is opposite, the one yielding a divergent model, the other (hopefully) a convergent one. These purely formal objections, by the way, especially that of the "crossing lines," fall on *all* the extramural uses of the complementarity principle I know of (e.g., in the social

sciences): they all are forced to violate the (at least) *semantically exclusionary* character the duality has in the original model and come to grief already on this count alone.

13. It is equally meaningful to ask *what* of our behavior, even of the mental state in back of it, is conditioned or circumscribed or prescribed by physical necessity, and what we truly initiate—i.e., to *apportion* the relative *shares* of the two sides in a given instance: again something wholly inapplicable to complementarity in the genuine sense.

14. For Spinoza, I have shown this in an article, "Parallelism and Complementarity: The Psychophysical Problem in Spinoza and in the Succession of Niels Bohr," in *The Philosophy of Baruch Spinoza,* ed. R. Kennington (Studies in Philosophy and the History of Philosophy, vol. 7) (Washington, D.C.: Catholic University of America Press, 1980), pp. 121–130 (title there mutilated by typesetter's error). The discussion of complementarity there is mostly identical with the present one.

15. E. Schrödinger, *Naturwissenschaften* 23 (1935): 807.

16. To the objection that no statistician would dream of making such a statement about person x on the basis of a life expectancy table for the population (but would say that no pronouncement at all is possible on the single case), the answer is that the two cases are not analogous. In normal statistics, the knowledge is about a population, and so, of course, are the predictions based on it. None on individuals are to be expected; but in "Schrödinger's cat," the initial knowledge is precisely of the state of the individual system (in the optimal case: as exhaustive a knowledge of "compatible" quantities defining it as on principle can be had together), and so predictions on the future state of *that* very system *are* to be expected and indeed are provided by the Schrödinger function. The proper analogy, therefore, is not between the cat and person x, but between the cat and the population to which x belongs— and there the prediction that two-thirds of "it" (namely, of its unspecified present members) will be alive and one-third dead at t_1 is perfectly meaningful and, maybe, true. But for the indivisible cat it is "true" only as a teaser. For both, of course, the simple and nonproblematical statement would do that the odds for the individual to be alive are two-thirds, and one might leave it at that, but then would blanket a profound difference. In the ordinary statistical case the statement merely expresses (e.g., to the insurance company) the individual's membership in the sample from which the probability ratio had been averaged and is in no sense a *causal* statement: no dynamic analysis of the "state" of either whole or part is involved. In the Schrödinger case the statement does express precisely the internal— and intrinsically probabilistic—"causality" of the analyzed individual system-state itself: a hardly comparable situation.

It is this difference, and the unorthodox nature of prediction under quantum-mechanical, probabilistic conditions as opposed to those of classical mechanics (under which population statistics are fully comprised) that the paradoxical—admittedly facetious—expression chosen by my interlocutor was to convey. It would be lost by submerging it in the classical relation of large numbers versus individual instances, which does yield precise (though not causally derived) predictions—for the large numbers. (This again is in reply to objections raised by Adolf Lowe.)

17. In population statistics we can allow a large measure of determinism concerning rates of mortality, births, crime, etc., without determining thereby any single cases, e.g., whether and when *I* die, procreate, commit a crime. The single cases are taken to be *diversely determined* by clusters of individual causes of their own—known or unknown, but anyway ignored—which average out in the large enough population sample. Note again the difference from the seemingly identical outcome in the Schrödinger example. There, to obtain that outcome, we had to stipulate for the parts initial conditions as nearly identical as they can be made. No such condition is imposed on ordinary statistics. There, on the contrary, the definite ratio results for the whole from the confluence of indefinitely *different* initial conditions (and consequent causal histories) in the parts. No such differences are averaged out in the multiplication of Schrödinger's cat, rather are identities converted from probabilities

in the parts toward certainty in the whole. The ratio coming out at the end was there in each of the parts at the beginning, and what is averaged out—toward determinacy—by the large number is merely the many, equal-valued indeterminacies, not many different determinations toward a mean value. Or, the mean value is given by the probabilistic equation of the single case. (In an analogous "classical" case, e.g., if I know of a population of insects that they all are ruled by the same biological clock, I do not, when I know its setting for one specimen, need statistics to tell me that the whole population—number irrelevant—will die at the onset of winter. But then, the individual biological clock is not probabilistic.)

18. This is technically not only feasible but quite familiar: any Geiger counter registering single radiation particles is an instance of it.

19. One attempt known to me from the camp of quantum physics, which avoids the temptation of complementarity and squarely tackles interaction, is L. Bass, "A Quantum Mechanical Mind-Body Interaction," *Foundations of Physics* 5/1 (1975): 159–172. It takes its cue from the introduction of the *consciousness* of an observer as an essential part of one version of the postulates of quantum mechanics (von Neumann), according to which one can say, "The impression which one gains at an interaction, called also the result of an observation, modifies the wave function of the system" (E. P. Wigner). In "a short step further," Bass imagines the inanimate microsystem thus modifiable to be placed in a strategic element of the central nervous system of the observer himself, so "that a suitable wave function pertains to a nerve cell which undergoes excitation when the wave function is modified by an event in consciousness." In this way, the connection between mind (= observer) and microscopic systems, posited in the general theory, is "brought to bear on the possible connection between the mind and the body of the observer" (p. 160). A formal model of such an element in the central nervous system is then constructed by Bass. Crucial for its working is, of course, that the element in question is "observed." Since, evidently, it is not so by the consciousness of the person, Bass resorts to the introduction of "subbrains," invoking C. S. Sherrington's view of the mind as "a collection of quasi-independent perceptual minds integrated in large measure by temporal concurrence of experience" (p. 170). Any such "subbrain," recording the relevant datum as a "quasi-independent mind," may then play the role of the "intermediate observer," and the "integrated mind" of the subject the role of the "ultimate observer" who records the state of the neural net *in toto*. Among quantum physicists this "ultimate observer" is known under the name "Wigner's friend," since it is borrowed from a thought experiment by E. P. Wigner.

The internal mathematical soundness of the model construction is beyond my judgment. I also waive the objection that we have no evidence of the existence of "subbrains"; and that anyway what we need would be "subminds"—by no means the same thing; and that to speak of mind parts, as one can perfectly well speak of brain parts, is questionable in itself, and the inference to a part-mind from a local brain part (even if this *were* a part-brain) is questionable once more: I waive this type of objection, because widest largesse shall be allowed to speculation at this stage. Not to be waived is the objection that the basic premise of the model construct confounds or conflates, under the blanket term of "consciousness," two different things: the state, or event, of subjective awareness and the objective act, or event, by which it was obtained, namely, measurement, which is a *physical* intervention in the state of the object. Here, Wigner's language—in the above quotation—is ambiguous. "The impression which one gains at an interaction . . . ": Is it the gaining or the having of the impression that modifies the wave function of the object? Surely it is the former, as a case of true physical interaction. Otherwise, the requirement of simultaneity is waived and we could even modify, as late observers, the wave functions of systems in the remote past, inasmuch as we gain knowledge of them via the intervening history of the universe. Nobody can mean that our ideas of the past can change the past. But so long as we stick to simultaneity of observer and observed and then place the modifying of the object in the *extramental* activity of *gaining* information, where by quantum-mechanical rights it belongs, we have not left the strictly physical sphere and still do not know how the mental datum once there, the information gained and held in consciousness, *then* in turn can become the fountainhead of a causal traffic with the object that is to be swayed *because* of (in answer to) "information received." Not what in the act of perceiving we may unknowingly have done to the object, but what, possessed of the percept, we may now or later knowingly do to the object—acting on it with our bodies and thus in the first place *making* our bodies so act: that is the causal hub of the mind-body problem. The initial puzzle still stares us in the face.

Index